Turret Mill Operation

John G. Edwards

HANSER GARDNER PUBLICATIONS

Hanser Gardner Publications
Cincinnati

Edwards, John G.
Turret Mill Operation / John G. Edwards
p. 21.6 cm × 27.95 cm
ISBN 1-56990-273-9
1. Turret milling machines. I. Title
TJ1225.E39 1998
621.9'1--dc21 99-14140
CIP

While the advice and information contained in this book are believed to be true, accurate, and reliable, neither the author or the publisher can accept any legal responsibility for any errors, omissions, or damages that may arise out of the use of this advice and information. The author and publisher make no warranty of any kind, expressed or implied, with regard to the material contained in this work.

Hanser Gardner Publications
6915 Valley Avenue
Cincinnati, OH 45244-3029

6 5 4 08 09

Table of Contents

Introduction

This book is written for the novice and machinist-in-training. Starting from a very basic level you will learn to build one chapter upon another and, with any luck, attain the skills required for basic vertical knee-type turret mill operation.

You will explore the nomenclature and operation of the vertical knee-type turret milling machine. Many of the tools and operations discussed can be utilized on other mills — both vertical and horizontal types.

Vertical knee-type turret milling machine operation is one of the most basic within the machine tool industry. Learning how to operate this type of milling machine is essential to your machine shop educational process. This machine is used for both production and prototype work alike.

Becoming acquainted with the skills necessary for mill operation will take some time. Do not become frustrated and give up; all skills are learned and refined over a long period of time. This book will not be able to give you all of the answers necessary to tackle each and every machining situation; however, you should be able to use your experience gained from the ensuing pages to at least get pretty close. You should, after working through this book, be able to gain the basic knowledge necessary to get started.

Machinists are *problem solvers*. Designers come to them with a set of drawings and need someone to transfer their ideas from a sheet of paper into a tangible part. By analyzing and studying the drawings, a machinist can produce the most complex of ideas and dimension lines into workable pieces, sometimes resembling works of art.

This book is laid out in chapters, each containing specific amounts of information. If you are new to the machinist trade, complete and understand each chapter in sequence. This will save you a lot of aggravation and shorten your learning curve. The first chapter will deal with specific issues pertaining to operator and machine safety. Chapter 2 will contain information about machine maintenance. Failure to maintain any machinery will shorten its life span and cause you to produce parts of low or substandard quality. Chapter 3 will delve into the history and features of the mill. Chapter 4 will contain a wealth of information about tool holders used to interface cutting tools with the machine.

Chapter 5 is all about the cutting tools used for milling operations. It will discuss the common tools that are commercially available from tool suppliers. Chapter 6 is focused on the tooling used to hold and position your workpiece. Loose or improperly positioned workpieces are a danger to you and the machine alike. Chapter 7 discusses tool cooling and lubrication, a necessary component for proper machine operation.

Chapter 8 centers on machine head alignment and how to perform this most basic of machine operations. In Chapter 9 you will master the skills required to move the table with handfeed and power-feed handles along with analog and digital movements.

Chapter 10 will prepare you to properly plan your jobs. The old Chinese proverb "A journey starts with the first step . . ." was never more appropriate than when you plan your job to include tools, holders, speeds and feeds, hold-downs—and the list goes on and on. Chapter 11 includes a series of projects designed to help you reinforce the materials learned in the previous chapters.

A small glossary of terms has been included to help you learn the language of the machinist. As with any profession, machinists speak a language that is particular to the trade and not often understood by the outsider.

You will find that use of the mill, like most machines, is a series of repetitious steps and setups. Each job performed on the mill will certainly use the same or similar steps as the one before. A workpiece must be held in place, aligned to the machine or tool, and then machined. Keep this in mind as you work your way through this handbook and the machine. The basic set-up steps are tools used to fill your bag of knowledge. Learning the actual steps required to turn a piece of raw stock into a finished part will take time, practice, and a lot of patience. Don't get frustrated and just walk away, use the tools in your bag of knowledge and you will become successful. Most machinists are made, not born.

About the Author . . .

Photograph by Bob Andre

John G. Edwards has been involved in the machinist trade, in one form or another, since he left high school and college. He has operated the majority of manual machines commonly found in most small and medium job shops. As so many hobbyists do, John was lured into the automotive business. First working for himself and then a car dealership, he never forgot his machinist roots. After the dealership failed he started another automotive repair business, but with a twist. Not only did he repair cars, he also started to make racing parts because he could not find the types and quality he wanted. Components included: pistons, lifters, pushrods, custom oil pans, headers, valves, and whatever else caught his fancy.

Soon after John was married he went to work for Keith Black Racing Engines as a fabricator. During that period he designed and built a variety of manifolds, both intake and exhaust, and other racing components. Finally the city, smog, and rat race got to John and his wife, so they moved to Ventura, California. There John was employed as an automotive machinist — again he never lost his machinist roots. In 1984 the local high school auto instructor wanted to go on sabbatical he thought John would be the perfect stand-in. This *stand-in* position grew into a full-time position. Currently, John teaches for the Ventura County Superintendent of Schools, Department of Secondary Education's Regional Occupational Program where he teaches automotive engine rebuilding and machine tool technology.

John started his writing career by writing the *Auto Dictionary*, published by HP Books. His other publications include: *The Yamaha 600 Air-Cooled Engine* motorcycle repair book, Haynes Publications; *The Small Scale Electric Vehicle Constructors Handbook*, JO/EL Engineering Technical Publications; and the just released *Sunnen's Complete Cylinder Head And Engine Rebuilding Handbook*, Sunnen Products Company.

Preface

The quality and completeness of any work is the sum total of those people, places, and entities who unselfishly give of their time and resources to make things happen. This publication demonstrates that premise to the highest degree. A great deal of effort has been provided to bring out the best of current thinking and industry-accepted practices from a community of friends, individuals, and machinists—a shared vision that we all have.

As already stated, this book is written for the novice and machinist-in-training. Everyone involved with this project has shared his or her years of experience to help shorten the learning curve. We don't want anyone coming away from reading the text to feel lost. Great effort has been expended to state operations and principles clearly. Many examples, photographs, charts, and illustrations have been included to help those *visual learners* to "get it."

I wish to thank the following individual people and companies for graciously giving of their time and materials to make this publication a reality. All of my students for urging me to write this book; AKKT Tool (Camarillo, CA); Bob Andre (Newbury Park, CA); Dick Bradbury (Bridgeport Machines, Inc.); the entire shop staff at Bridgeport Machines, Inc.; Woody Chapman (Hanser-Gardner Publications); Joseph E. Clancy (Bridgeport Machine, Inc.); Bob Cook (Bridgeport Machines, Inc.); Maryann Czarnecki (Bridgeport Machines, Inc.); Randy Buchanan (Buchanan Precision Machine); Javier Garcia (Barbers Machine); the entire staff at Hanser-Gardner Publications; Dan Griffith (Bridgeport Machine, Inc.); Kenny Helderlein (H.E.C., Inc.); the entire staff at Helderlein Engineering Consultants (Oxnard, CA); Image Source (Ventura, CA); JO/EL Engineering Technical Publications (Ventura, CA); Ralph LoStocco (Bridgeport Machines, Inc.); Jack Luper (J. Luper Machine); the Magana brothers—Mark and LaRoy; Photo Communications, Inc. (New Haven, CT); Scott Rabe (Ventura College); Roy Transfiguration, H.E.C.; special thanks to Scott Turnbull (Bridgeport Machines, Inc.); the Ventura County Superintendent of Schools' Regional Occupational Program; Peggy Velarde, principal of the Regional Occupational Program; and John Van Zant (my best friend and mentor). And, as always, my wife Elaine—thanks again, honey pie, you're the best!

Finally, a very special *THANK YOU* to Bridgeport Machine, Inc. who made this book possible. Those mentioned above made my pilgrimage to Connecticut unforgettable. Not only was I given full freedom to wander about the plant, I was also allowed to get to know the plant staff. Their dedication to their profession reminded me of another era and time. Perhaps this is because a good number of the workers and technicians were European trained, and some of them are second or third generation Bridgeport employees. The pride they take in their work is truly inspirational, and I want to express my gratitude and admiration to and for Bridgeport Machine, Inc., and rethank everyone who endeavored to make my stay a memorable and unique experience.

Chapter 1
Safety

Safety in the shop is important, very important. You will not be able to perform the duties of a good machinist very well with only one eye, or missing fingers and limbs.

***Figure 1.1** Use proper safety equipment when operating your mill. This machinist is wearing snug-fitting pants, a short-sleeved shirt, approved eye protection, and steel-toed boots. Notice the lack of jewelry and watch, as well.*

Working around machinery requires a high level of awareness of your surroundings and operational procedures. Using your senses to see, hear, and feel what you do may save untold pain and suffering. Keep your eyes on what you are doing. Listen for unusual sounds. Also, learn to feel and distinguish good from bad vibrations.

Look . . .

You are only issued one set of eyes when you're born. You must learn to take care of them. Wear eye protection whenever you work around tools and equipment. Your eyes are the most delicate part of your exterior body. They are sensitive to flying objects like chips and broken tools.

Eye protection takes several forms: face shields, goggles, and glasses. Each has a place within the machine shop.

- Face shields use a large piece of thin plastic sheeting attached to an adjustable headband to keep the large pieces of debris from colliding with your face – a nice feature when using a flycutter or drill.

***Figure 1.2** The face shield provides a minimal amount of protection, but does protect your entire face.*

- Goggles are those protective lenses that use a flexible plastic frame and a stretch elastic band to keep them in place. Goggles are best used around grinders where they provide complete protection for the eyes from flying abrasive grit and metal shavings.

***Figure 1.3** Goggles provide the most protection because they completely surround your eyes.*

- Glasses provide the most convenient form of eye protection and are, by far, the most comfortable. There are a myriad of frame types, including: all plastic, metal frames with shatter-resistant plastic lenses, and metal frames with shatter-resistant glass lenses. The majority of safety glasses will incorporate some type of side protection.

Figure 1.4 *Glasses are by far the most popular form of eye protection used thoughout the machine shop industry.*

Most machine shops and schools will have a minimum eye protection requirement of glasses with side impact protection. The majority of safety glasses will feature built-in side shields. If you wear glasses to correct vision deficiencies, most shops may allow you to use them for your safety glasses. However, you may have to purchase some sort of side-impact shields to bring them up to the minimum shop requirements. If you wear glass lens glasses, there is a hardening process that you may want to employ to increase shatter resistance. Wear your safety glasses at all times when you are in the shop area.

Your hands and fingers are also very important. The rules here are very simple – keep them away from moving parts. It may seem very inviting to reach in and pick away chips from a rotating end mill. Don't even think about reaching for a chip. Just when you don't expect it, something else will jump out at you. Your skin makes a poor substitute for a chip raker.

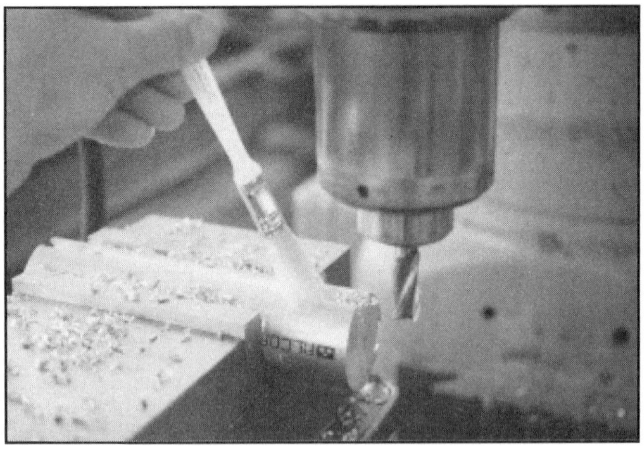

Figure 1.5 *DON'T use your hands and fingers to remove chips from the workpiece during machining!*

If you must remove chips from a machine in motion, use an air nozzle, or brush. For wiry chips that wind themselves around end mills and drill bits, stop the machine to remove them. This doesn't mean to grab them as the spindle slows down – no – let it come to a full and complete stop.

Safety shields are installed onto machines for your protection, not the machine. Never remove them unless you put them back in place.

Listen . . .

Learn to use your ears as you work. You can't move the table, or turn the machine on and off with your ears, but you can learn to distinguish which sounds are definitely out of the ordinary.

Figure 1.6 *Be constantly aware of your surroundings, learn to distinguish what is normal and typical background shop noise. And pay attention to what you're doing. Don't let your mind wander, this will lead to an accident for sure.*

A dry running end mill will produce a slightly different pitch than one that is run with coolant. Dull cutting tools are also easily identified by their cutting sound. The pitch of the machining process can also distinguish fast or slow feed rates.

Figure 1.7 *Use approved hearing protection when machining certain difficult materials, or when sound pressure levels exceed 90 decibels.*

There will be occasions when you will need to wear hearing protectors. Certain materials will create a high pitched sound during machining which can be very detrimental to your hearing. In this case it is not a reflection upon the machinist's inability to select the correct tool, the material is just that difficult to machine. In cases where hearing protection must be worn, be extra careful when listening to shop background noise.

Feel . . .

By placing your hand on the mill table you will learn how to detect the results from good and bad vibrations. There is a close relationship between your hearing and feeling senses. This is not to say that you should keep one hand on the mill and the other on the feed handle, not at all. You may want to place your hand onto the table if you start to hear something out of the ordinary. Or, you may want to feel the table when you start the machining of a part to ensure that it is tightened down properly.

Figure 1.8 *Always feel the machine table when you start machining your workpiece. If you feel vibrations from the vise and not the table, stop the machining process and find out why.*

And, In the End . . .

In the end, you are responsible for your own personal safety in the shop. Shop owners and schools alike can only provide you with the knowledge and expectations necessary for safe machine operation. It's up to you, the individual, to work safely. Removing existing or user-installed safety equipment can only lead to system malfunctions and injury to the machine operator. Injuries caused by machine failures are very rare indeed. Tool failures, on the other hand, are common and caused by the operator. Excessive *pushing* or *crowding* of the tool will induce or cause failure. Learning how to compute proper speeds and feeds is essential to ensure long and expected tool life. Exploding tool bits not only cause personal injury, but can be quite costly in terms of tooling and downtime.

There is a saying, "Guns don't kill people, people kill people." The same can be said of the mill in its relationship to accidents. A mill at rest tends to stay at rest, unless some sort of user interface is applied. The bottom line – don't do stupid things and blame it on the machine. Your safety is most important. Take the time to learn the safety precautions supplied by your mill manufacturer and use your own best common sense.

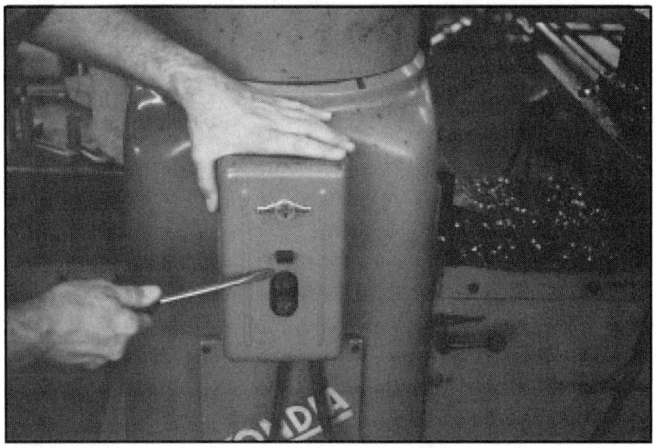

Figure 1.9 *Don't fool with the motor controls or try to override their functions. You may expose yourself to great personal injury, or death. Always call qualified personnel to work on machine electrical systems.*

The mill is heavy. It should be moved and set into place only by experienced people. Do not try to move the mill by yourself unless you are familiar with safe equipment moving practices.

Accessories used on the mill may be heavy. A typical 6" mill vise may weigh around 90 pounds (40.9 kilograms), or more. Storage of vises and moving them around the mill will require some thought. Don't try to lift a heavy vise all by yourself. Use safe lifting practices. Use your arms and legs when lifting – not your back. That means to hold the part close to your body and use your arms and legs as the primary lifting force. Holding a heavy object at an arms length away from your body locates the pressure at acute angles and will cause back strain. Again, get help when necessary.

Figure 1.10 *Always get help, or use a cart, when moving heavy objects around the shop – your back will thank you.*

When working around certain materials you may also want to wear a dust protector to keep small particles of debris out of your lungs. Certain materials may also *out-gas*. This is a phenomenon that occurs during machining as the waste material heats up and liberates a gaseous byproduct.

Figure 1.11 *Use appropriate air filtration masks when necessary. Your health may depend upon it!*

The following 36 tips are for your personal safety . . .

1. DON'T run your machine until you have read and understood all operating instructions.
2. DON'T run your machine until you have read and understood all machine and control key signs.
3. DON'T run your machine for the first time without a qualified instructor, ask him or her for help when you need it.
4. PROTECT your eyes. Wear safety glasses with side shields at all times.
5. DON'T get caught in moving parts. Remove watches, rings, jewelry, neck-ties, and loose fitting clothes.
6. PROTECT your head. Wear a safety helmet when working near overhead hazards.
7. KEEP your hair away from moving parts.
8. PROTECT your feet. Always wear safety shoes with steel toes and oil resistant soles.
9. Gloves are easily caught in moving parts. TAKE THEM OFF before your turn on the machine.
10. Loose objects can become flying particles. REMOVE all loose items (wrenches, chuck keys, rags, etc.) from the machine before starting.
11. NEVER operate a machine after taking strong medication, using non-prescription drugs, or consuming alcoholic beverages.
12. SAFEGUARD the cutting zone (*point of operation*). Use standard, general purpose safeguards where possible. Use special safeguards when required.
13. PROTECT your hands. STOP the spindle completely BEFORE changing tools.
14. PROTECT your hands. STOP the spindle completely BEFORE you load or unload a workpiece.
15. PROTECT your hands. STOP the spindle completely BEFORE you clear away chips or oil. Use a brush or chip scraper. NEVER use your hands or fingers.
16. PROTECT your hands. STOP the spindle completely BEFORE you adjust the workpiece, fixture, or coolant nozzle.
17. PROTECT your hands. STOP the spindle completely BEFORE you take measurements.
18. PROTECT your hands. STOP the spindle completely BEFORE you open safeguards or covers.
19. NEVER reach around a safeguard.
20. PROTECT your hands. STOP the machine BEFORE you change or adjust belts, pulleys, or gears.
21. PROTECT your hands. Keep hands and arms clear of spindle start switch when changing tools.
22. PROTECT your eyes and the machine. NEVER USE A COMPRESSED AIR HOSE TO REMOVE CHIPS.
23. KEEP work area well-lighted. Ask for additional light if needed.
24. DON'T slip. Keep your work area clean and dry. Remove chips, oil, and obstacles.
25. NEVER lean on your machine. Stand away when the machine is running.
26. DON'T get trapped. Avoid pinch points caused by motion of the table and head.
27. PREVENT objects from flying loose. Securely clamp and locate the workpiece. Use stop blocks where necessary. KEEP clamps clear of cutter path.
28. PREVENT cutter breakage. Use correct table feed and spindle speed for the job. Reduce feed and speed if you notice unusual noise or vibration.
29. PREVENT cutter breakage. Rotate spindle in clockwise (CW) direction for right-hand tools, and counterclockwise (CCW) for left-hand tools. Use the correct tool for the job.
30. PREVENT workpiece and cutter damage. Never start the machine when the cutter is in contact with the workpiece.
31. KEEP tools sharp. Dull or damaged tools break easily. Inspect tools and tool holders. Keep tool overhang short.
32. KEEP rotating cranks and hand wheels well-lubricated and maintained. Do not remove safety guards.
33. Certain materials, such as magnesium, are highly flammable in dust and chip form. See your instructor or supervisor before working on such materials.
34. PREVENT fire. Keep flammable liquids and materials away from work area and hot chips.
35. PREVENT machine table from moving unexpectedly. Disengage power feed when not being used.
36. PREVENT machine from moving unexpectedly. Always start machine in manual mode.

There is no end to safety, and the same is true in this book. You will find safety precautions and suggestions throughout. Pay attention to all of them, they are not here to just *fill* the book with more pages. Your personal safety is paramount to everyone in this business, and most of all to yourself.

Next . . .

In the next chapter we will look at the maintenance of the mill. Daily maintenance is essential to keep your machine in top condition. We will examine the daily and routine maintenance of the turret mill.

Notes:

Chapter 1 ◆ Study Questions

1. Machinist A says that safety is the responsibility of everybody in the shop, Machinist B says that attention should be paid to shop background noise. Who is correct?
   A. Machinist A
   B. Machinist B
   C. Both A and B
   D. Neither A nor B

2. Machinist A says that you should feel the mill table when you are nearly through with the initial cut. Machinist B says that you should listen to the cutter as you machine the workpiece. Who is correct?
   A. Machinist A
   B. Machinist B
   C. Both A and B
   D. Neither A nor B

3. Machinist A says that dry running tool bits produce a different sound than those run with coolant, Machinist B says that the pitch of the sound may be different. Who is correct?
   A. Machinist A
   B. Machinist B
   C. Both A and B
   D. Neither A nor B

4. What should you wear on your head when working around overhead hazards?
   A. Hair net
   B. Baseball cap
   C. Baseball hard hat
   D. None of these

5. At what point should you wear hearing protection?
   A. 60 dB
   B. 70 dB
   C. 80 dB
   D. 90 dB

6. Pushing a tool bit is recommended when machining . . .
   A. Tool steels
   B. All aluminum alloys
   C. Chrome alloy steels
   D. It is not recommended

7. Care should be taken when . . .
   A. Operating any machinery
   B. Removing tooling from the machine
   C. Removing parts from the machine
   D. All of the above

8. Machinist A says that you should wear gloves when operating a milling machine, Machinist B says that you should remove gloves when operating a milling machine. Who is correct?
   A. Machinist A
   B. Machinist B
   C. Both A and B
   D. Neither A nor B

9. Machinist A says that vises may be very heavy, Machinist B says that you should lift heavy objects with your legs – not your back. Who is correct?
   A. Machinist A
   B. Machinist B
   C. Both A and B
   D. Neither A nor B

Chapter 2
Maintenance

Lubrication

Like all machines, mills require periodic and routine maintenance. Lubrication and cleaning will comprise the majority of your maintenance duties.

Figure 2.1 *Use a shop vacuum and brushes to remove unwanted debris from the mill. Be sure to sweep up your mess when you are through and dispose of the unwanted material properly.*

Keeping your machine lubricated is essential to long life and good service. The mill should be completely lubricated daily and, in most cases more often. Machine lubrication schedules are dependent upon the number of hours the machine is used every day. Every mill manufacturer publishes a recommended lubrication interval chart.

Figure 2.2 *Machine one-shot oilers should be used every day to keep the mill in tip-top shape. Follow the machine manufacturer's guideline for the grade and type of oil used for way and bearing lubrication purposes.*

The head requires lubrication via an oiling point near the front side. By lifting the spring-loaded cover you can access the oiling orifice. Pour the correct oil into the fill-hole. Once the oil reaches the top of the filling port, stop the filling process. Spring-loaded oilers will close automatically. If the machine has screw-top oiling ports, unscrew the cap, lubricate, and return the cap.

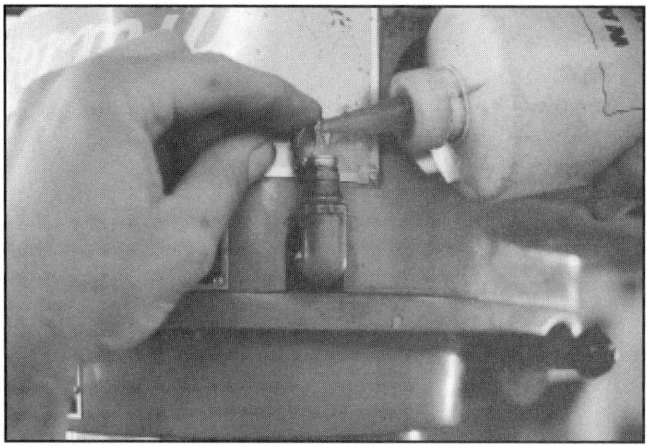

Figure 2.3 *Be sure to lubricate the spindle and motor oilers daily!*

The majority of mills will use some sort of *one-shot lubrication* system to lubricate the table and knee. This type of lubricator utilizes an oil reservoir and pump. Filling the reservoir is accomplished by removing the fill cap and filling it with the correct weight *way oil*. The machine is lubricated when the pump handle is activated through a pumping motion. Typically, two to three pumps or strokes of the handle is all that is required to lubricate the machine. Machines in use for six or more hours per day should be lubricated at least twice daily.

Figure 2.4 *Use a brush to apply lubrication to the machine table and knee lead screws.*

A light coating of medium to heavy way oil should be placed onto all of the machined surfaces of the mill. This will help to keep rust and the effects of corrosion to a minimum.

Figure 2.5 *After thoroughly cleaning the machine, apply a liberal coating of oil to the machine ways and a light coating to the machine table to prevent rust. Dispose of the oily rag(s) properly.*

The following lubricants are recommended for machine way lubrication:

- ◆ Gulfway 68
- ◆ Sunoco Way Lubricant 1180
- ◆ Mobil Vactra Oil No. 2
- ◆ Way Lubricant 68 (Texaco)
- ◆ Tonna 68 (Shell)

Do not over-lubricate the quill. The quill is sensitive to dust and particulate matter accumulation if excessive amounts of lubrication are used. The contamination may cause the quill to wear rapidly as it is moved in and out of the machine head.

Cleaning

Keep your mill clean. Cleaning is always a pain and nobody really likes doing it – but clean we must. The accuracy of the mill is paramount if we are going to produce acceptable parts. Maintain a daily routine for cleaning, and the job is not so bad. However, if you let up, for even one day, the whole clean-up process becomes torture. Small metal chips can work their way into the smallest of nooks and crannies and, if left there, will start to grind (a loose term here) grooves into the machine ways. This is a situation that compounds itself when the machine is not properly lubricated.

Figure 2.6 *Clean the mill on a daily basis.*

The most common means used for clean up in the shop is compressed air. It doesn't require a lot of work from the operator and it quickly removes debris. However, the use of compressed air for cleaning purposes on the mill is discouraged, instead use a wet or dry shop vacuum.

Excessive air pressure can blow small chips into places where they can't ever be removed. Never use more than 30 psi (2.0 BAR) air pressure for cleaning purposes. All that you really want to do is to sweep away the chips and debris, not launch them into orbit. Remember . . . always wear eye protection when using any compressed air or machine tool.

Never point the end of the air nozzle at anyone! And, never point it at yourself. Compressed air, when blown into a small cut, can separate skin from the muscle tissue. This is very painful and can cause an *air embolism*. Your heart is a marvel at pumping liquids, however, it was not designed to pump air. Air in the heart or the brain can cause certain death. Be careful with compressed air.

Figure 2.7 *Never set the air pressure above 30-psi (2 BAR) when cleaning the top of the mill table. Always wear approved eye protection when using compressed air.*

For cleaning the T-slots there are several solutions. The most common is the T-slot cleaning key. It has the same shape of the T-slot and removes the vast majority of chips in one pass. A pocket is located at each end of the T-slots on the table. It has three functions: 1) to collect the chips removed from the T-slots; 2) to facilitate T-slot tooling installation; and 3) to provide a pool for excess lubrication and coolant fluids. Tapped holes on the backside of the table allow the mounting of plumbing for coolant recirculation. Screens are necessary to keep the larger chips out of the coolant flow. Check the screens periodically for clogging, and clean as necessary.

Figure 2.8 *Use a T-slot key to clean the majority of debris from T-slots.*

When finished with a job you will want to clean the machine, especially if coolant was used during the machining process. Excess mineral (oil) based coolants should be scraped into the machine T-slots if possible. Oil splashed onto the machine's other surface areas may require the use of paper, or cloth, towels and a spray cleaner/degreaser. Dispose of paper and cloth towels properly. When using water based coolants, be sure to wipe all surfaces after you are finished machining. This does not mean wait for several days after the job is through. Water based coolants are very corrosive and can cause very rapid wear to the machine. Again, dispose of the rags and towels properly.

Coolant tanks for flood-type cooling will require periodic maintenance. Follow coolant manufacturer's recommendations for handling and disposal procedures.

The installation of way protectors is a very good idea. They keep the majority of chips away from the machined way surfaces. Clean the protectors with a *fox-tail* brush.

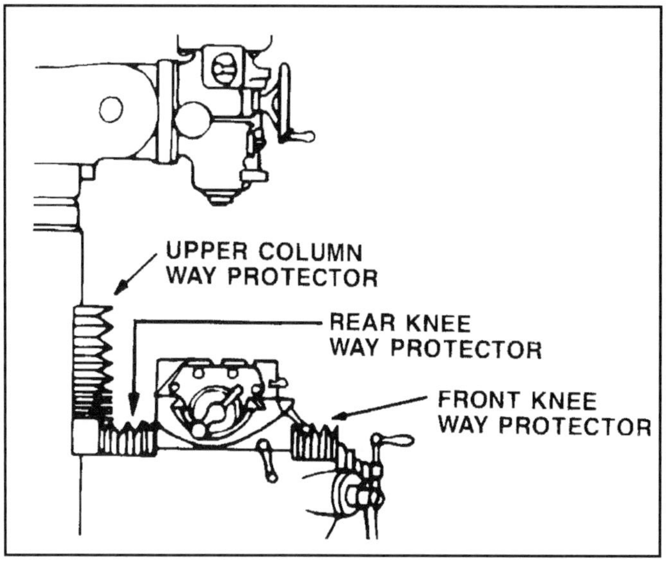

Figure 2.9 *Plastic way protectors reduce clean up time and provide good protection for the machine ways.*

Major exterior cleaning of the mill should be performed at least once a week, or more often if the machine is used heavily, or when using heavy cutting oils. The use of mineral spirits or an aqueous solution is most effective for cleaning the painted surfaces. All cleaners are designed to *degrease* the surface of the mill or part. By removing the grease, or oil, you open up the pores of the machine for oxidation (rust). Always lubricate any machined surfaces with way oil after cleaning.

Figure 2.10 *The painted surfaces of the machine should be cleaned with approved cleaning solutions. Wear protective clothing as recommended by the chemical manufacturer.*

If the machined surfaces require heavy cleaning to remove sludge or dried-out coolant, use a plastic scraper — not steel. The machined surfaces must not be scratched during cleaning. A small gouge in the surface can create a pocket for tiny chips to hide and accelerate the wear between any rubbing surface.

Also, keep the area around the mill clean. Sweep chips into a pile and dispose of them properly. Mop the area with an approved cleaning agent to remove oil and grease spots. *Slip* and *fall* hazards are one of the leading contributors to lost productivity.

Wear appropriate clothing and hand protection when cleaning any machine. Remember that chips may puncture certain types of rubber gloves. Use approved respirators when using those cleaners that may cause personal respiratory or health problems.

> *Never flood the floor with cleaning solution and touch the machine at the same time. An electrical shock hazard may exist. Electric shock can be fatal to humans and animals.*

Description of Maintenance Procedures

1. Check lube oil system oil level; fill if necessary. The most convenient procedure is to keep the level of oil to the top of the tank. Check it every day.
2. Clear dust and chips from the ways at the end of the day. Use a brush and/or an industrial wet or dry vacuum cleaner, then wipe carefully to remove damage-causing abrasive material. Do not use compressed air to clean the ways. **Note:** Do not use compressed air to clean the ways or around the cabinets. Using compressed air could blow chips and other foreign material into the interlocking parts, control system, or at the operator, resulting in extensive damage or serious injury.
3. Clean machine exterior, clean intake and exhausts. Clear dirt and chips from machine at the end of the day. Use an industrial wet or dry vacuum cleaner, then wipe carefully to remove damage-causing abrasive material. Do not use compressed air to clean the machine.
4. Clean and apply a light coat of oil to the way covers once a week to keep them pliable.
5. Check coolant level; fill as necessary.
   Flood Coolant: Keep hose joint areas free of chips and dirt. The coolant will come out of the nozzle in spurts when level is too low. Fill with comparable antibacterial emulsified cutting oil.
   Mist Coolant: Wipe off excess grime from the top of the coolant reservoir periodically. It is recommended filling it only with the amount to be used in one operation.
6. Check pneumatic regulator system bowls; fill lubricator bowl if necessary. When the level drops below EMPTY line, fill to FULL with suitable lubricant, e.g., Mobil DTE 24, or equivalent.
   a. Shut off air pressure.
   b. Unscrew the screw from the fill hole from lubricator.
   c. Fill bowl to the FULL line. Do not overfill.
   d. Replace the fill hole screw.
7. Drain and clear refill pneumatic regulator bowls. The bowl should be drained whenever it fills up with sludge. The bowl should be cleaned semiannually.
   a. Put a dry rag under the regulator to catch the drips.
   b. Drain the filter bowl by pressing up on the drain valve.
   To clean the bowl:
   a. Drain the filter bowl.
   b. Gently unscrew both bowls.
   c. Rinse them with WARM WATER only. Do not use soap. **Note:** Use only warm water to clean the filter bowls. Using soap, solvent, or chemicals may weaken the bowl and cause it to burst, resulting in serious personal injury.
   d. Replace the bowls. Fill the lubricator with suitable lubricating oil, e.g., Mobil DTE 24, or direct equivalent.
8. Check the spindle motor for dirt; wipe if necessary. The spindle motor can become overheated if excessive grease and dirt are allowed to build up on it.
   a. Remove spindle motor hood.
   b. Inspect the motor for dirt.
   c. Wipe the motor with a rag and remove as much of the build-up as possible.
   d. Replace the cover.
9. Check spindle drive belt for dirt and wear. If the housing itself is excessively dirty, the belt may be worn or weak. Have belt replaced by qualified service technician.
   a. Remove the spindle drive belt cover.
   b. Inspect the belt for wear, cracks, or damage.
   c. If the belt looks worn, have belt replaced by qualified technician.
   d. Replace cover.

Check each item listed in the checklist (in your appropriate Operator Daily Maintenance Requirements Table) on a regular basis.

Periodic Codes:

D = Daily, performed on a daily basis during a 24 hour period.
W = Weekly, performed on a weekly basis.
M = Monthly, performed on a monthly basis.
S = Semiannual, performed twice a year.
AS REQ = Depends upon work environment.

Ref #	Requirement	Periodicity				
		D	W	M	S	As Req.
1	Check lube system oil level; fill if necessary	X				
2	Clean dirt, chips from ways	X				
3	Clean machine exterior; clear intakes and exhausts		X			
4	Clean way covers; lightly oil		X			
5	Check coolant level; fill if necessary		X			
6	Check pneumatic lubricator bowl; fill if necessary			X		
7	Drain, clean, refill pneumatic filter bowl. See warnings.				X	
8	Check spindle motor for dirt; wipe if necessary			X		
9	Check spindle drive belt for wear; notify Dealer Service to replace if necessary			X		
10	Clean lube system pump filter				X	

Figure 2.11 *The Operator Maintenance Requirements checklist.*

Next . . .

In the next chapter you will learn about the vertical knee-type turret mill features and controls. There are many features, and understanding their operation is very important.

Notes:

Chapter 2 ◆ Study Questions

1. Machinist A says that the mill should be lubricated at least once a week, Machinist B says that the one-shot lubricators are installed on the mill head. Who is correct?
   A. Machinist A
   B. Machinist B
   C. Both A and B
   D. Neither A nor B

2. Machinist A says that the quill should be lubricated whenever the mill is used, Machinist B says that the ways should also be lubricated. Who is correct?
   A. Machinist A
   B. Machinist B
   C. Both A and B
   D. Neither A nor B

3. Machinist A says that 30 weight non-detergent motor oil should be used to lubricate the knee and ways, Machinist B says that way oil should be used to lubricate the quill. Who is correct?
   A. Machinist A
   B. Machinist B
   C. Both A and B
   D. Neither A nor B

4. Machinist A says that compressed air can be used to remove chips from the T-slots, Machinist B says that you should wear eye protection when using an air nozzle. Who is correct?
   A. Machinist A
   B. Machinist B
   C. Both A and B
   D. Neither A nor B

5. Way protectors are used to . . .
   A. Keep debris out of the spindle
   B. Keep debris out of the quill feed
   C. Keep debris out of the table ways
   D. None of these

6. Machinist A says that degreasers are used to remove oil and grime from the surface of the machine, Machinist B says that the machined surfaces of the machine should never be coated with oil to prevent gum and varnish formation. Who is correct?
   A. Machinist A
   B. Machinist B
   C. Both A and B
   D. Neither A nor B

7. Machinist A says to use a T-slot key to clean heavy dirt and grease deposits from the machine ways, Machinist B says to use a plastic scraper on the table. Who is correct?
   A. Machinist A
   B. Machinist B
   C. Both A and B
   D. Neither A nor B

8. Machinist A says that small scratches in the ways allow the pooling of lubricating oil which is good for lubrication, Machinist B says that a clean work area is a safe work area. Who is correct?
   A. Machinist A
   B. Machinist B
   C. Both A and B
   D. Neither A nor B

9. Machinist A says to check the screens on the table for blockage when using mist coolers, Machinist B says that the screens are used to clean the machine lubricating oil administered by the one-shot oiler. Who is correct?
   A. Machinist A
   B. Machinist B
   C. Both A and B
   D. Neither A nor B

10. Machinist A says to wear appropriate clothing when cleaning the mill, Machinist B says to use a respirator if necessary. Who is correct?
    A. Machinist A
    B. Machinist B
    C. Both A and B
    D. Neither A nor B

Chapter 3
Knee-Type Turret Mills

The vertical knee-type turret mill is a rather imposing piece of machinery. It stands taller than the average-height man or woman and weighs well over one ton (900 kilos). This machine is considered to be one of the backbones of the machine shop industry. The machine is very flexible and allows the use of a great number of options for part holding and machining parts.

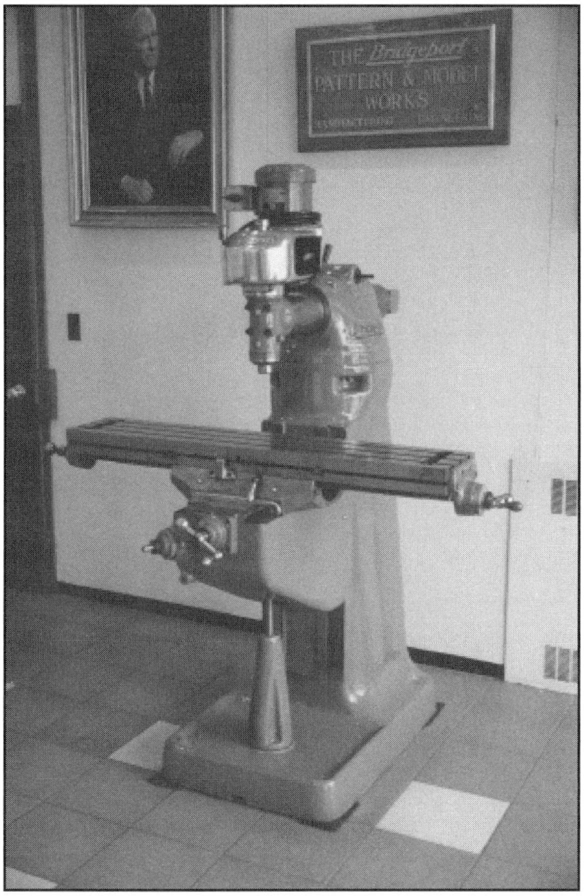

Figure 3.1 *This is an example of the very first turret mill ever produced (serial number 7).*

The Knee-type turret mill has been in use since 1939. The basic shape and size has not changed drastically since that time. The first units produced utilized a round ram (often referred to as the *J* head) instead of the familiar V-ram type (often referred to as the *2J* head) we see today. The round ram, or column, carried a 3/4 or 1 horsepower motor to power the mill head. To improve *side-load* stability, the V-ram was designed. After many years of use the round column would rotate from the side loads (X-axis operation) placed upon it during long production runs if not clamped securely.

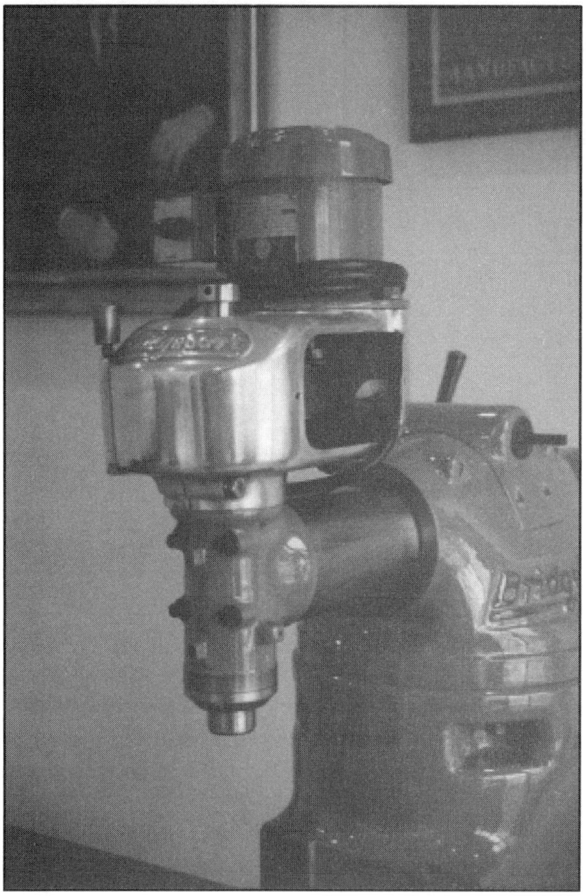

Figure 3.2 *The first mill was a round over-arm-type knee milling machine.*

Today's mill is available with a variety of factory, and user, installed accessories, including: power feed units for the table (X- and Y-axis) and knee (Z-axis); variable speed heads; one-shot lubrication systems; power draw bars; chrome ways; oversize (length and width) tables; digital readouts; and the list goes on. (See Figure 3.3.)

With an eye toward increased production and accuracy, many shops are retrofitting special *stepper* motors to the machine feed handles and coupling them to a computer controller. These converted machines can now function as a low cost *computer numerical control* (CNC) machine. These converted machines must be used within a set of predefined and practical limits because of the machine's size and construction. Compared to purpose-built CNCs, the V-type ram mill is *light duty*, both in performance and weight. (See Figure 3.4.)

Figure 3.3 This mill is fitted with a power table feed and digital readout.

Figure 3.4 This mill has been "factory-fitted" with a set of special motors and a controller. It can be programmed in a "teach mode" by moving the table to a designated location and recording the information, or by "up-loading" information created on a computer.

Computer assisted mills are the perfect machine for shops and schools because of their versatility and low cost. Students and trained machinists both find the machine easy to learn and operate.

Illustrations 3.5, 3.6, 3.7, and 3.8 provide different views of the machine.

Figure 3.5 Right front view.

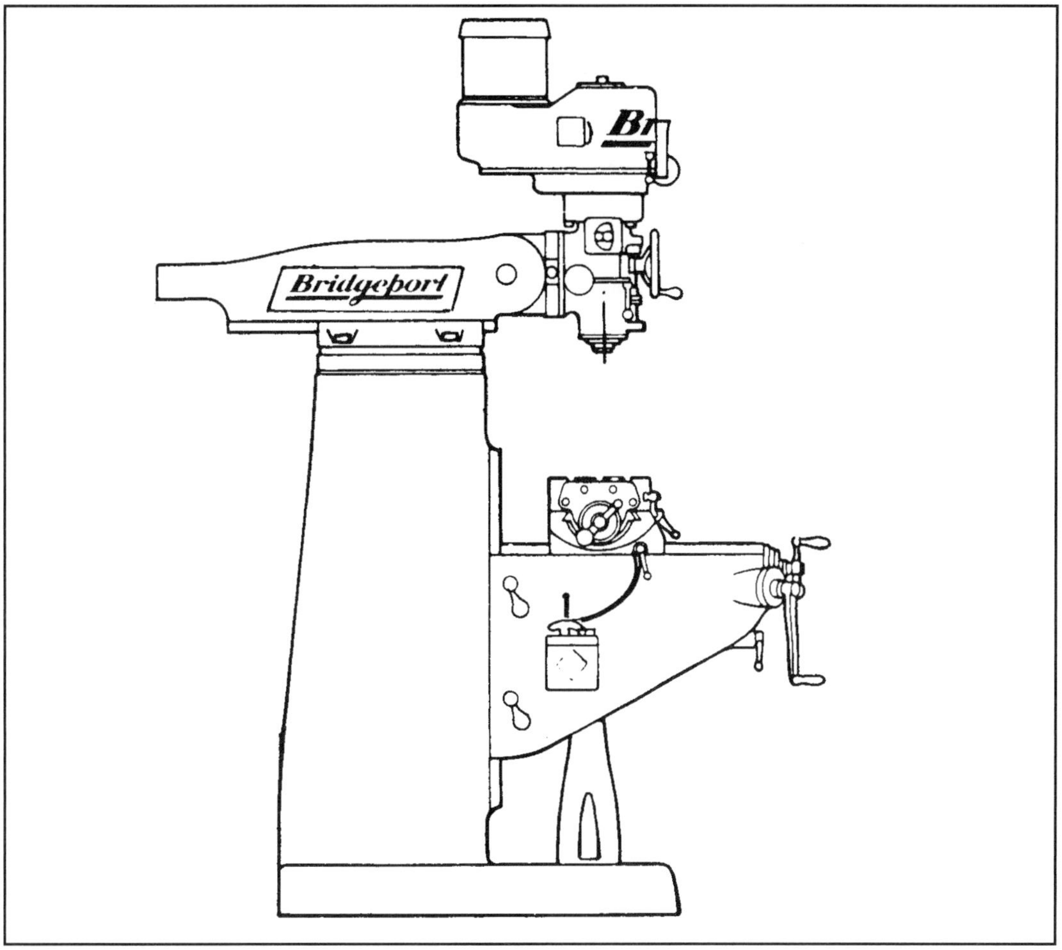

Figure 3.6 Left side view.

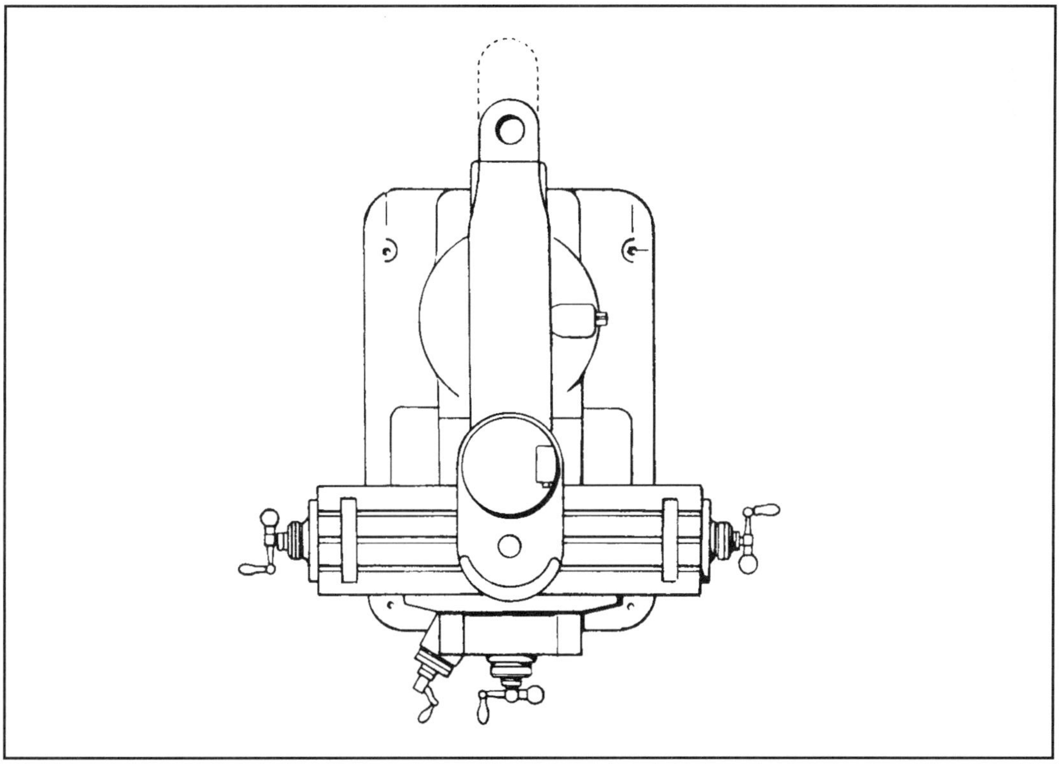

Figure 3.7 Top view.

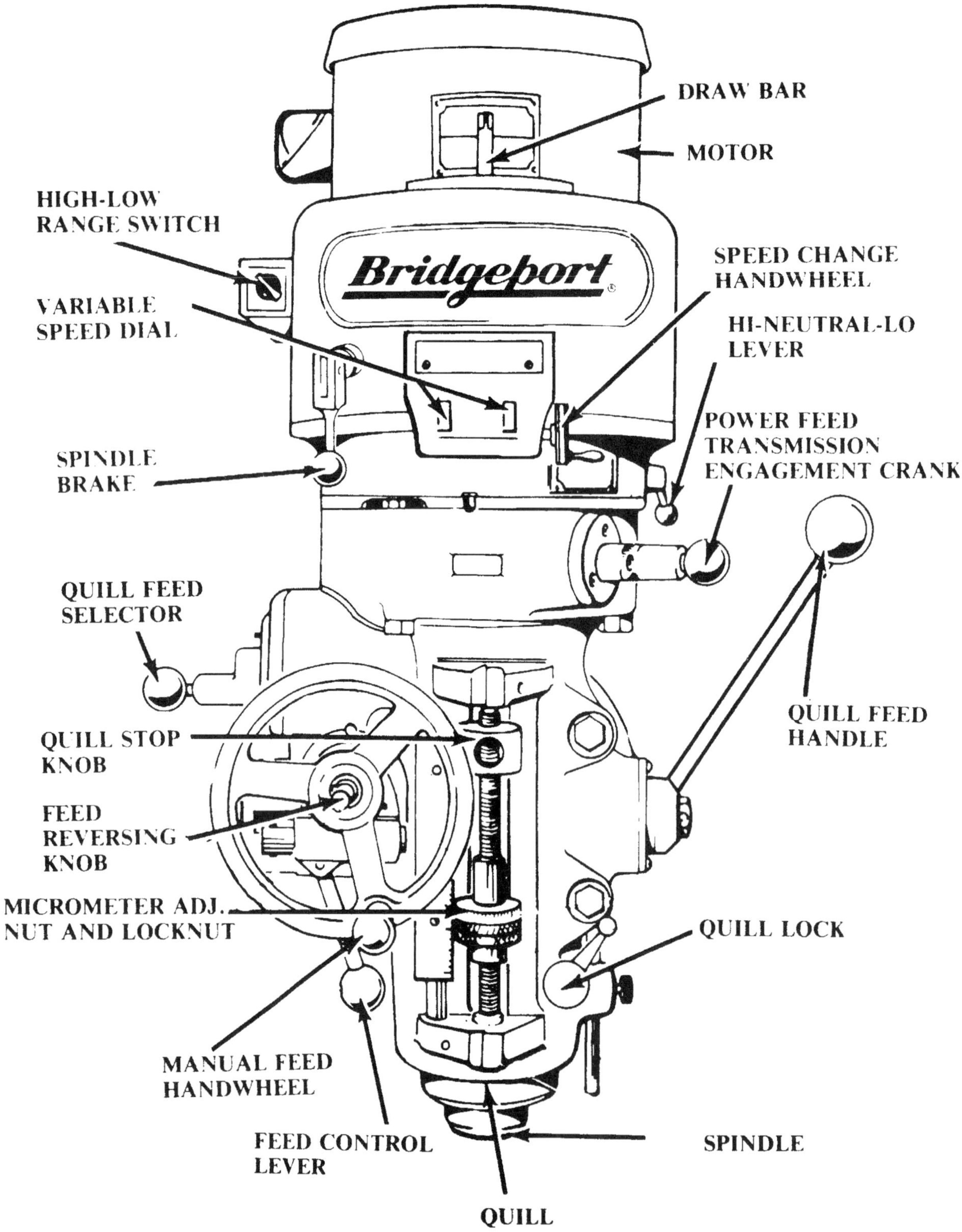

Figure 3.8 *Head Controls, w/variable speed control.*

Electrical Controls

All of the electrical controls are designed with operator ease and safety in mind. The controls are located within easy reach of the operator to reduce fatigue and increase productivity. There are a variety of switches and levers that you will need to become familiar with in order to operate the machine with ease.

Figure 3.9 *The machine electrical controls are positioned for ease of use and simplicity — a Forward and Reverse switch.*

Mills are typically *soft-wired* to the shop power supply (typically 220- or 440-VAC 3ϕ). This means that the power cord is covered with an oil and abrasion resistant covering in a continuous run from the wall to the machine. This allows the mill to be moved as needed. Turret mills must be considered *portable machines*. You may have to move the mill from one position to another from time to time to allow long parts of machine stock to hang over the edge of the table. Remember, you will need to re-level the mill every time you move it.

Local electrical codes may require your machine to have a remote magnetic relay switch. This switch may be mounted on the wall or on the side of the machine. The relay switch is a safety device installed for the protection of the operator. Do not tamper with any electrical switches or try to circumvent their operation. Failure to do so may result in personal injury or death.

Figure 3.10 *Certain areas and locales may require that machines be fitted with a remote magnetic switch. These switches and circuit breakers should be turned OFF when making any repairs to the machine. Always tag the circuit breaker to let others in the shop know that the machine is under repair and that the circuit should not be turned ON.*

The main control for the spindle is the *Forward-Reverse* switch. It will normally be located on the mill head. Depending upon the switch supplier, the words *Forward* and *Reverse*, *1* and *2*, or *High* and *Low*, will be imprinted on the switch box. This two-way switch is used to direct the spindle rotation – forward (clockwise, CW), or reverse (counter-clockwise, CCW). **Note:** The head must be set in the *high-speed* range. When the spindle is placed in the low-speed range (sometimes called the *back-gear*), the direction of the spindle will reverse in relationship to the posted spindle direction. This means that Forward becomes Reverse and Reverse becomes Forward. Always check the spindle rotation before you start machining your workpiece. Also, cutters can be purchased in *right-hand* (RH) and *left-hand* (LH) rotations. This will affect the actual rotation of the spindle. Turning a cutter bit backwards will cause the tool to fail very quickly.

Figure 3.11 *The top gear shift lever on manual speed change machines must be set to correspond with either the high or low speed ranges.*

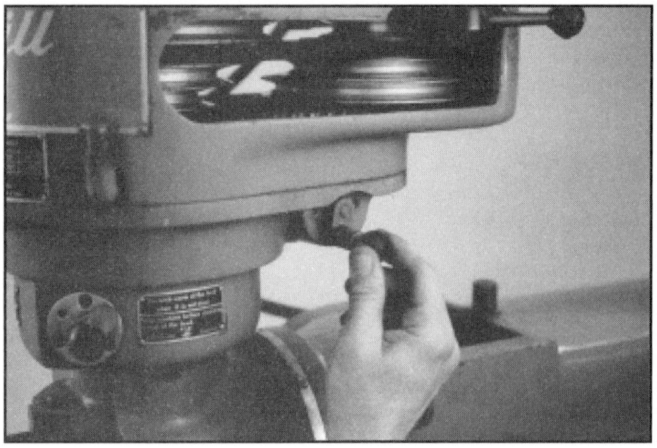

Figure 3.12 *The side gear lever must also be adjusted to correspond to the correct high or low speed ranges.*

The most frequent addition to any mill will be that of a table power feed unit. They can be installed at the factory, or in the field. The long (X-axis) feed of the table is the most likely recipient of a power feed unit. Power feed units provide the operator with a steady and variable feed that cannot be easily duplicated by manual operation. Power feeds improve the surface finish of the part in a very dramatic fashion. Cross-feed (Y-axis) and knee (Z-axis) power feed units are also available. The knee unit is especially handy in production shops where the table must be raised and lowered frequently.

Figure 3.13 *This machine is fitted with a table power feed unit. Table direction and speed are controlled with the lever (direction) and rheostat knob (speed).*

Power feed units use a rheostat to control their speed. A lever is used to switch the feed on and off. A button switch is used for rapid traverse of the table or knee. This is a handy feature for getting from one location to another very quickly. A pair of stop rods are located in the front edge of the table which, when adjusted, can limit the power feed travel of the table. They also act as an automatic safety shut-off in case the operator is detained from turning the unit off.

Figure 3.14 *The power feed shut-off switch is used to turn OFF the power when the table reaches predefined limits. The stop-buttons can be set within a certain range to allow the feed operation to be performed while the machine is unattended — the table will stop moving, but the spindle will remain ON.*

Spindle speeds are changed manually, or with a variable speed changer. Manual machines can be set to eight different speeds: four high speed, and four low speed. The variable speed head, like the manual head, has two speed ranges – low and high – but they can be set to any desirable speed within the high and low maximum and minimum.

Step-Pulley (J-Style) Head (Manual Machines)

Manual speed changes are made by changing the belt positions on a *stepped pulley* system — one on the motor and the other on the spindle. This is how the machine gets its name — *stepped head.* In order to change speeds the machine must be turned off.

To change from one speed to another (high or low range), use the following procedure.

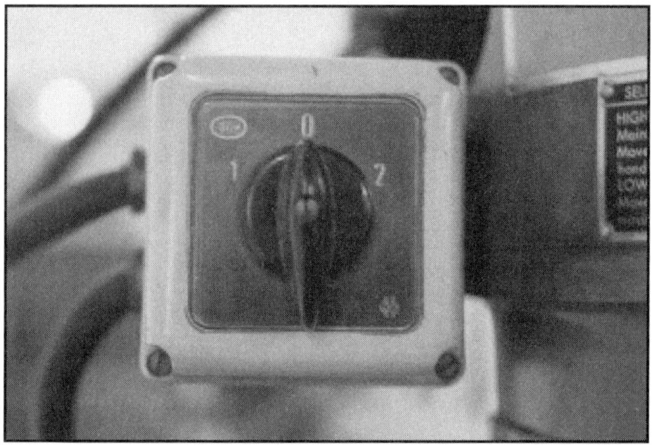

1. Ensure that the spindle motor is turned **OFF**.

2. Loosen the motor clamp bolt which secures the motor to the head.

3. Pull the motor forward. This will cause the drive belt to become slack.

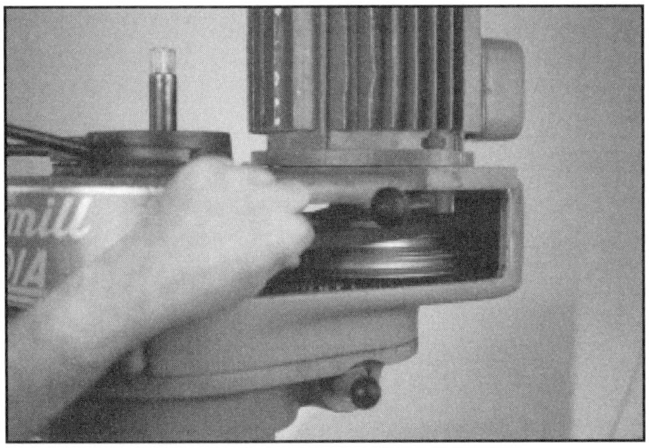

4. Move the belt to the desired speed position.

5. Push the motor towards the back of the machine and hold in place.

6. Tighten the motor clamp bolt and you are ready to go.

When the speed must be reduced or increased to a level where the simple belt change will not help, you will need to shift the machine from high to low, or from the low to high speed ranges. This change must be made with great care in order that the machine is not damaged. The following steps are used for this procedure.

High-to-Low Speed Change (Into Back Gear)

1. Ensure that the spindle motor is turned **OFF**.

2. Pull the top spindle lever to the right side of the machine, approximately 45°.

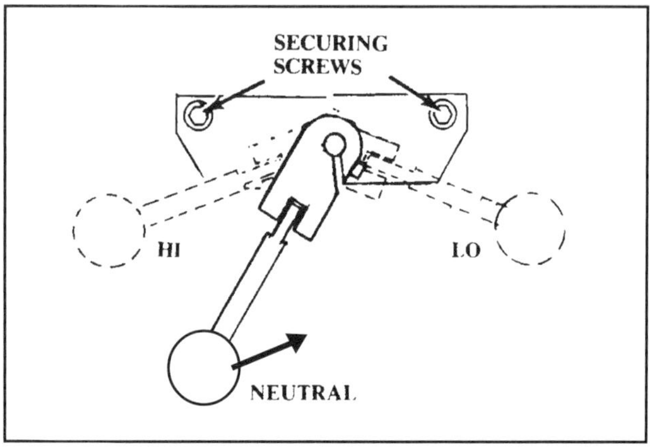

3. Rotate the right side speed control lever approximately 180° to the "low" position. Some rotation of the spindle may be necessary to fully engage the spindle.

Low-to-High Speed Change (Out Of Back Gear)

1. Ensure that the spindle motor is turned **OFF**.

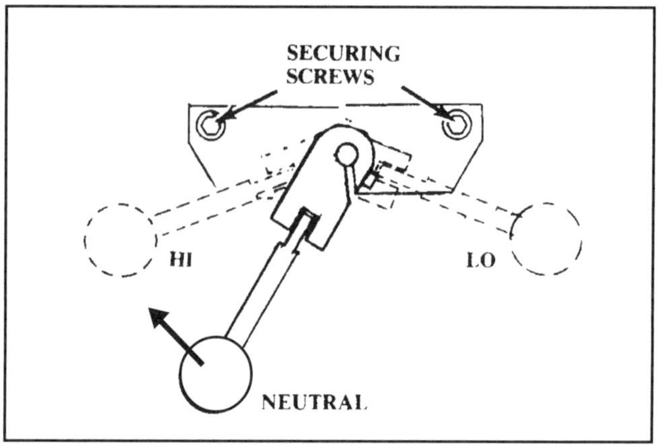

2. Rotate the right side speed control lever approximately 180° to the "high" position. Some rotation of the spindle may be necessary to fully engage the spindle.

3. Pull the top speed control lever to the left side of the machine, approximately 45°.

4. Rotate the spindle until an audible *click* is heard. **Note:** Failure to lock the spindle at this point may cause catastrophic damage to the mill head.

> The spindle speed change will necessitate the ON-OFF switch to change directions as well. Forward will become reverse, reverse will change to forward.

Variable Speed (2J-Style) Head

Speed change on the variable speed machines is very simple compared to the stepped head types. This head also has two speed ranges — high and low. However, the means by which they are changed is different than previously discussed.

The following steps are used to change speeds on the variable speed head.

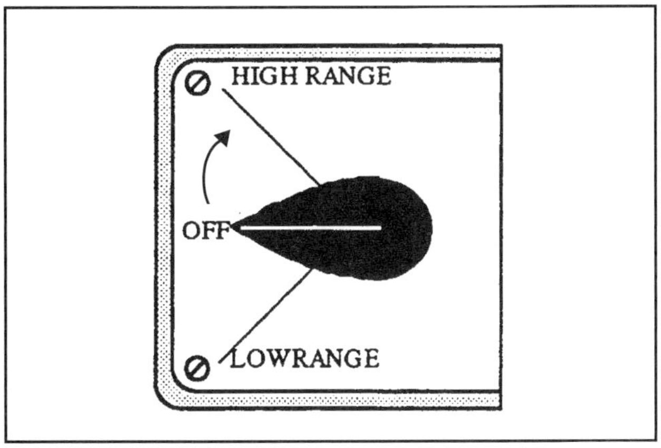

1. Turn the machine **ON**. All changes on the variable speed head must be made with the spindle on or serious damage may occur to the mill head.

2. Rotate the speed change hand wheel located on the right side of the mill head until the desired speed is obtained.

High-to-Low Speed Change (Into Back Gear)

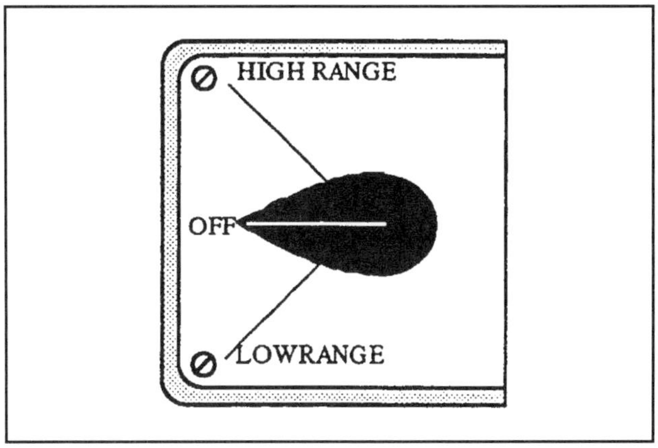

1. Turn the machine **OFF**.

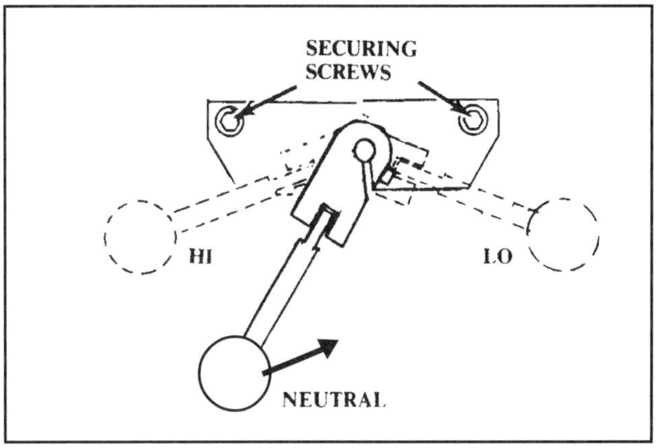

2. Rotate the right side speed control lever approximately 180° to the "low" position. Some rotation of the spindle may be necessary to fully engage the spindle.

Low-to-High Speed Change (Into Back Gear)

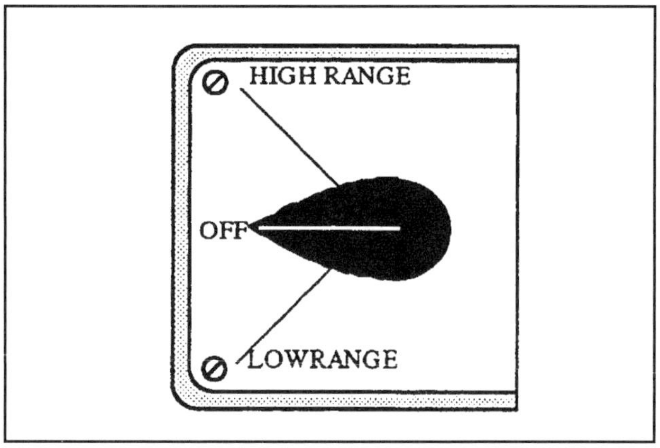

1. Ensure that the machine power is turned **OFF**.

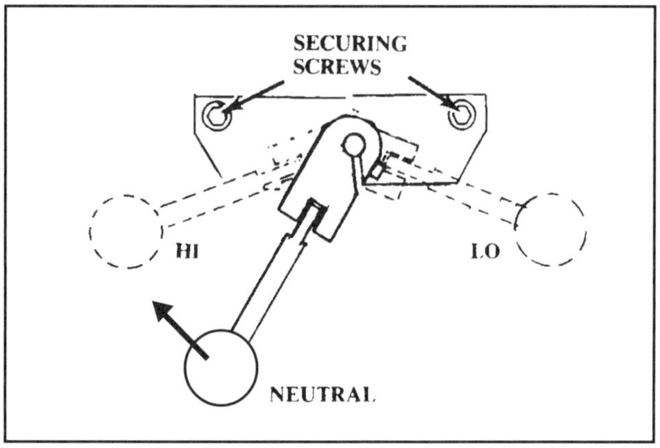

2. Rotate the right side speed control lever approximately 180° to the "high" position. Some hand rotation of the spindle may be necessary to fully engage the spindle.

Spindle/Quill Feed

Moving the quill moves the spindle in and out of the power head. The main quill feed is located on the right side of the power head — a movable handle with a round ball at the end. The handle must be able to be repositioned to suit the exact location of the spindle. This makes for less operator fatigue and easier machine operation. For special operations and fine feed adjustments you will also find a round feed wheel on the front of the power head. This feed wheel is used when very precise movements of the quill are necessary. The round feed wheel can be removed for storage when not in use. The power ON/OFF lever on the right side of the head can also be used to disconnect the fine feed from the spindle power feed.

Figure 3.15 *The quill is moved with the long lever handle on the right side of the machine head which can easily be removed (shown) and repositioned to make operation easier.*

During production runs, the spindle will be used to machine, or bore, holes to specified depths time after time. In order to repeat the process with a degree of accuracy you will need to adjust the spindle depth. This is done with the spindle depth control located on the front of the machine. There are two nuts located on a threaded rod. The top nut has numbers that correspond to depth in thousandths (0.000") of an inch. One revolution of the nut will equal .050". This nut is sometimes referred to as the *micrometer stop*. The lower nut is used as a lock or jam nut to keep the top nut from moving out of position. This *stop* feature is very handy and will be used many times.

Spindle Power Feed

Anytime you use the quill feed handle you are, in effect, acting like a power feed unit — you are providing the motive power to move the cutting tool. Self-power is not bad, but by having the machine perform this task we get a much better surface finish and less operator fatigue. Power spindle feeds are geared to the spindle speed. Each revolution of the spindle will move the quill a specified amount. This means that the feeds are listed as inches/millimeters per revolution. Typically, machines will move the quill .0015", .003", or .0045" per spindle revolution. The gear change lever is located on the left side of the machine head. A lever for engaging the power feed is located on the right side of the power head. This lever is used to engage/disengage the power feed from the spindle drive — not to engage the feed unit. The power down-feed is a very handy machine feature.

Figure 3.16 The power spindle feed is activated with the control lever on the left side of the machine head.

It is most important to ensure that all cutting tools have sufficient support. Less than ideal support will result in tapered holes if spindly tool bits are used to machine a hole. This would be the case when boring a hole with an adjustable boring head fitted with a small diameter tool bit. The chance of making a tapered hole straight is nearly impossible once a hole becomes tapered.

Figure 3.17 Feed rate changes are made by pulling the lever ball out and repositioning the lever to the desired feed rate. Be sure that the lever is correctly engaged before you turn ON the power.

Table

The mill table is precision machined and mated specifically to each machine. There are three T-slots (2-$^1/_2$" centers) machined into the surface of the table to facilitate workpiece clamping. Never hammer on the mill table or use it for any purpose than what it was designed for. The width and length of the table vary from machine manufacturer to manufacturer. Typical width will range from 9" - 11" wide, with lengths as short as 32", and as long as 60".

Ways

In order to keep the machine table straight and parallel to the machines knee a set of *ways* are used. The ways are typically a *dove tail* or *square* box-like arrangement. There are ways on the X- and Y-axis of the table and the knee. If square ways are used on the X- and Y-axis of the table, the knee will use a dovetail. An adjustable *gibb* is used to reduce side-play in the table as it is moved from one direction to another. Gibbs will require periodic adjustment as the machine starts to wear. In an effort to increase machine life the ways are chrome plated. This option is necessary when machines are expected to machine fibrous materials, or when installing automated table positioners.

Figure 3.18 "Dove tail" ways.

Figure 3.19 "Box" or "square" ways.

Each one of the ways on your machine is mated and matched at the factory. This means that you cannot use the knee and/or table from one machine base on another. To ensure that the ways are perfect, a process known as *scraping* is performed. Scraping is performed with a scraper and is done by hand. This means that each machine is somewhat unique to itself.

Figure 3.20 *The dovetail way on this knee cross slide are being "scraped-in" to ensure an exact fit with the table and base ways. This process is also used to align the table and knee to the mill base.*

Lubrication of the ways is performed with a *one-shot* lubricator. A pump/reservoir unit is mounted to the side of the knee and uses a series of tubes and fittings to lubricate the machine ways. The ways should be lubricated at least once a day for normal machine operation, every four hours under heavy usage.

Figure 3.21 *The "one-shot" lubricator should be used to lubricate the machine ways at least once or twice a day to ensure proper lubrication and operation.*

Keep the machine ways clean. Stray chips that may get stuck in the ways can cause premature wear and eventually affect the accuracy of the machine. To help prevent wear, some machines have a set of way protectors or felt wiping pads installed at the factory.

Figure 3.22 *Felt wiping pads are used to keep the ways clean and lubricated. Be sure to apply way oil lubricant to the felt wipers as part of your daily maintenance routine.*

Next . . .

In the next chapter we will take a look at the various types of tool holders used to interface cutting tools to the machine. As would be expected, there are a great many different holders used to suit the needs of various machine operations and procedures.

Notes:

Chapter 3 ◆ Study Questions

1. Machinist A says that 2J milling head was an improvement over the round ram type, Machinist B says that V-type ram mill is a light duty machine. Who is correct?
   A. Machinist A
   B. Machinist B
   C. Both A and B
   D. Neither A nor B

2. Machinist A says that most mills are wired for 220- or 440-VAC 3ϕ, Machinist B says that the main electrical switch is used to raise and lower the table knee. Who is correct?
   A. Machinist A
   B. Machinist B
   C. Both A and B
   D. Neither A nor B

3. Machinist A says that the magnetic relay switch is installed for production reasons, Machinist B says that the magnetic switch is installed for safety reasons? Who is correct?
   A. Machinist A
   B. Machinist B
   C. Both A and B
   D. Neither A nor B

4. Machinist A says that turning the main power switch to the right will turn the spindle clockwise (CW), Machinist B says that it depends on whether or not the machine is in back gear. Who is correct?
   A. Machinist A
   B. Machinist B
   C. Both A and B
   D. Neither A nor B

5. Machinist A says that power spindle feed is engaged with the *feed control* lever, Machinist B says that power feed transmission engagement crank must be disengaged before the power is applied. Who is correct?
   A. Machinist A
   B. Machinist B
   C. Both A and B
   D. Neither A nor B

6. Machinist A says that J head mill has variable spindle speeds, Machinist B says that you must adjust the belt positions to one of four pulleys to change speed on the variable speed machines. Who is correct?
   A. Machinist A
   B. Machinist B
   C. Both A and B
   D. Neither A nor B

7. Machinist A says that it is important to fully engage the spindle top gear before you start the machine once you have changed from the *Low* to the *High* speed range, Machinist B says that also remember to shift the *Hi-Neutral-Low* lever to the correct position when changing speeds. Who is correct?
   A. Machinist A
   B. Machinist B
   C. Both A and B
   D. Neither A nor B

8. Machinist A says that the spindle should be completely stopped before you change the spindle speed on variable speed machines, Machinist B says that you should use your hand to slow the spindle because it is faster than using the machine brake. Who is correct?
   A. Machinist A
   B. Machinist B
   C. Both A and B
   D. Neither A nor B

9. Machinist A says that turret mills are constructed with *box ways*, Machinist B says that they are constructed with *dove tail* ways. Who is correct?
   A. Machinist A
   B. Machinist B
   C. Both A and B
   D. Neither A nor B

10. Machinist A says that the *one-shot* lubrication system provides lubrication to all parts of the mill, Machinist B says that the head must be lubricated separately. Who is correct?
    A. Machinist A
    B. Machinist B
    C. Both A and B
    D. Neither A nor B

Chapter 4
Tool Holders

In order to machine a workpiece we must install a cutting tool into the machine. There are a variety of tool holding devices used to grip the cutting tool, and a more detailed explanation of cutting tools used to remove material will be discussed in the next chapter. It is necessary to understand the differences between tool holder designs before the proper selection can begin.

We will examine several of the most popular tool holding devices *commercially available*. The idea of "commercially available" is important to the designer and machinist alike. The cost to produce special tools, or make adaptations to standard tooling, can be quite substantial. Special tool fabrication also requires advance *lead times* in order to have the tool on hand when it's needed for a job. There are times, however, when special tools must be constructed. These tools will often increase production, which is desirable.

There are three types of tool systems used to hold the cutting tool: the R-8 (most popular), the #30 taper, and the 200 Series. These tool holders do not interchange with one another. The R-8 system uses a lightweight holder, sometimes called a *collet*. This makes for quick and easy tool changes and they provide the precision necessary to produce parts. Number 30 and 200 Series taper tooling are used where high production rates are expected. These holders are heavy and very rigid, qualities necessary for high production rates. Because the R-8 system is most commonly used, we will examine it in detail.

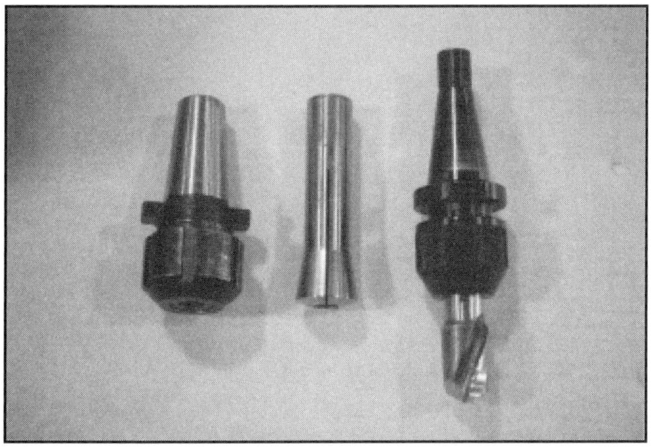

Figure 4.1 *The R-8 collet in the middle is the most commonly used type on turret-type milling machines. The #30 taper (right) and 200 Series (left) tool holders are used on high production machines and where powered drawbars are standard equipment.*

The popularity of the R-8 system has provided machinists with a large group of configurations for almost any possible machining scenario. Even though there is so much variety, most of the tooling has become what we call *standard tooling*. Collets are made in various standard sizes ranging from $^1/_{16}$" to $^3/_4$" diameters (in $^1/_{16}$" increments) in the SAE format, and 3 millimeters to 19 millimeters (in 1.0 mm increments) in the ISO format. Larger collets are available, but their holding and gripping ability is limited because the sidewalls are very thin.

Installation and alignment of the R-8 tool holder into the machine must be done with a skilled touch. A $^1/_8$" wide groove cut parallel to the collet length must be aligned with an internal alignment pin in the machine spindle. Failure to align the holder with the alignment pin may cause the pin to be sheared off, or cause the collet to spin when tightened.

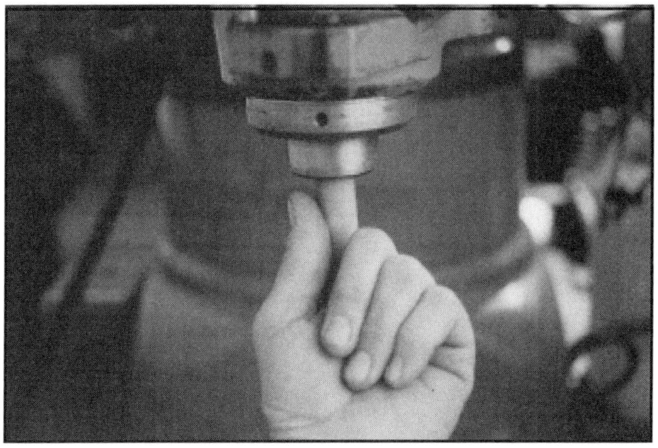

Figure 4.2 *Be sure that the alignment pin in the quill is aligned properly with the collet before it is installed.*

Tools designed for the R-8 system will have the same construction and dimensions necessary to align the tool accurately to the spindle centerline.

You need to understand the importance of proper tool support. When mounting cutters in an R-8, or any other tool holder, be sure that the tool is clamped properly. This means to place the cutter shank all the way into the collet or holder. Holding only $^1/_2$ or $^2/_3$ of the shank is a formula for disaster. Again, think SAFETY. It's always better to err on the side of caution.

Drawbars

The majority of standard tool holders (R-8 and #30 types) are held in place with a *drawbar*. Tightening the bar pulls the tool holder or collet into a seated position. The 10° taper cut onto the end of the R-8 collet squeezes against the tool as the drawbar is tightened. The more the drawbar is tightened, the more firmly the tool is held by the spindle.

Figure 4.3 *The drawbar is located on the top of the machine head. A $^{3}/_{4}$" wrench is used to tighten and loosen it. Note how this tool has a $^{3}/_{4}$" box-end at one end and plastic coated mallet on the other.*

Drawbars are inserted into the top of the machine head, then pass down and through the machine spindle. The R-8 drawbar has a $^{7}/_{16}$"-20 threaded end. At the opposite end close of the drawbar there is a $^{3}/_{4}$" hex bolt used to tighten the bar. If the drawbar is tightened by hand, the machine is said to have a *manual* drawbar system. Do not over tighten the drawbar, or you may strip the threads. This is a common mistake made by novice machinists.

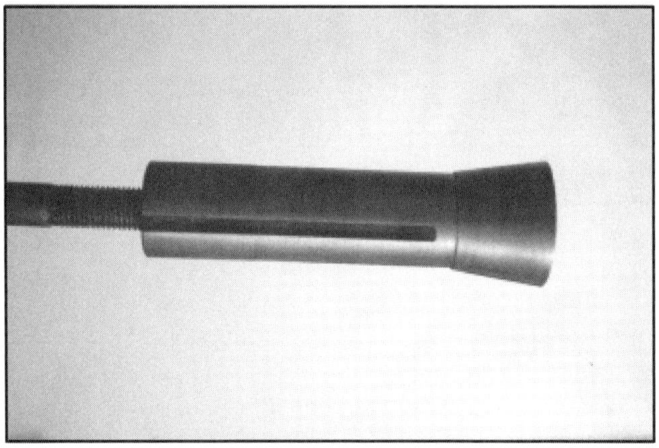

Figure 4.4 *The standard R-8 drawbar is $^{7}/_{16}$" in diameter and has 20 threads per inch (TPI).*

Powered *drawbars* are used to speed the tool change process. Power is typically provided via a pneumatic drive tool actuated by a control button or switch. Powered drawbars are convenient and are designed to decrease operator fatigue, and increase production.

Figure 4.5 *This machine is fitted with a pneumatic power drawbar.*

Figure 4.6 *Notice the UP and DOWN buttons used to tighten (UP) and loosen (DOWN) the collet.*

R-8 Collets

A typical set of R-8 SAE size collets would include the following sizes: $^{3}/_{16}$", $^{1}/_{4}$", $^{5}/_{16}$", $^{3}/_{8}$", $^{7}/_{16}$", $^{1}/_{2}$", $^{5}/_{8}$", and $^{3}/_{4}$". Sizes are available in increments of $^{1}/_{16}$" up to and including $^{7}/_{8}$" diameters. ISO collets are found in whole millimeter increments from 4 mm to 21 mm.

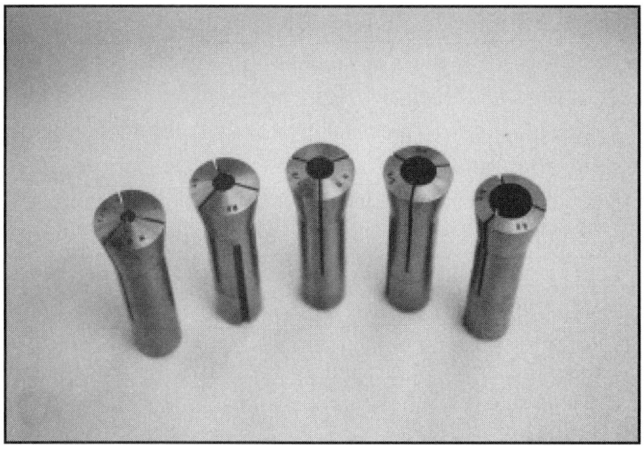

Figure 4.7 *A typical set of R-8 collets, $^1/_8$" to $^3/_4$" range.*

Split-Collet Holders

When a collet will not open or close sufficiently, another means of tool holding is necessary. Enter the *split-collet holder*. This group of tool holders uses a changeable double angled split-collet which is held in place with an R-8 mount. The split-collet allows more flexibility and size range than normal R-8 collets. They are typically used to hold drill bits because they are manufactured in $^1/_{32}$" and $^1/_{64}$" size increments. This is very helpful when holding drills with diameters slightly larger or smaller than the standard $^1/_{16}$" increment size R-8 collets will permit. Number and letter drills do not typically fall within the normal collet sizes available. Trying to use standard collets to hold slightly larger or smaller bits can decrease the tool's gripping power.

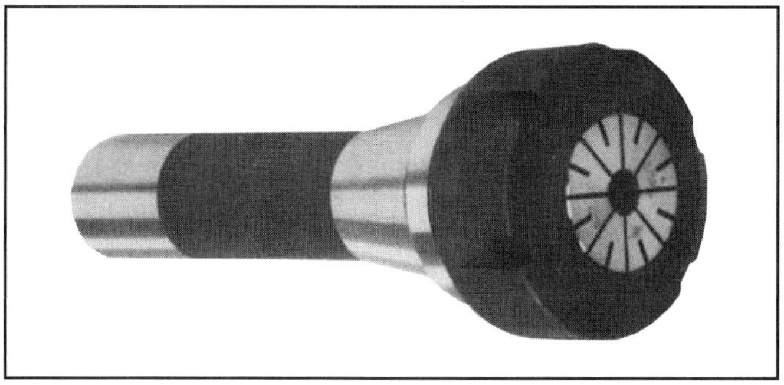

Figure 4.8 *Split-collets are very handy for odd-sized tool bits.*

End-Mill Holders

End-mill holders are very different from collets and collet holders. They are designed to hold a cutting tool with more rigidity than collet-type holders can offer. Tool bits are held in place with a set screw. Tool bits held with these holders typically have a *Weldon* shank, a machined-in relief for the set screw to seat itself. This type of shank is very desirable; otherwise the set screw will damage the outside diameter of the tool bit, making it difficult to remove and reinstall.

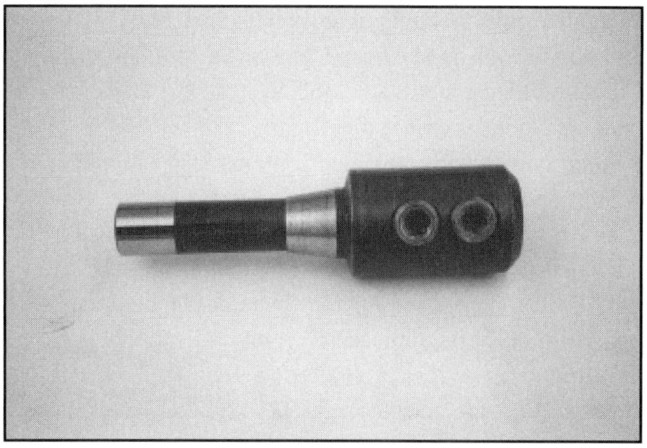

Figure 4.9 *End-mill holders provide very rigid support for mill tooling. The larger sizes allow you to hold tools with diameters over $^7/_8$" with ease.*

End-mill holders come in a variety of sizes ranging from $^1/_8$" to 1-$^1/_2$" diameter. They are perfect for holding tool bits larger than $^3/_4$" in diameter, and/or where extra rigidity is required.

Quick-Change Tooling

In an effort to increase the speed by which we turn out a product and improve the bottom line, *quick-change* tooling was invented. Some systems uses a master holder with an R-8 shank, and a series of rapid and interchangeable tool bit holders. The master holder must be made with great precision to insure that each interchangeable tool holder will seat and align correctly. The 200 Series quick change tool system uses a special spindle to hold the tool holders. Tooling is changed by rotating a special nut on the end of the spindle. Hand tightening is all that is required to change the tools from one to another.

Tool holders are typically supplied with normal cutter body sizes associated with standard collets, ranging in sizes from $^1/_8$" to $^5/_8$" diameter. Larger diameters are not typically used because of the increased amount of material required to fully support the tool bit.

Figure 4.10 *The use of quick-change tool holders is increasing rapidly among machinists. Not only can tools be changed more quickly, but also their tool length offsets (TLO) remain the same.*

Quick-change tool holders are easy to operate. Typically, they require only $^1/_2$ to $^3/_4$ of a turn to release the tool. The difference in time to change a tool with this system, compared with the typical drawbar system, is significant.

Some quick-change tooling systems also include special adapters and specialty tools to round out their system line. Adjustable boring heads can be altered quite easily using adapters, and specially adapted flycutters can also increase production speed. Quick-change tooling may become standard equipment in the near future and replace the typical collet holding systems we currently use.

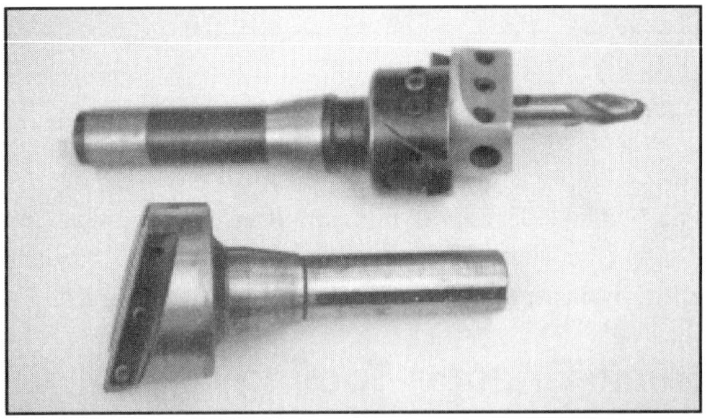

Figure 4.11 *Special tool holders can be used to mount unique tooling like this boring head and flycutter.*

You can create your own quick-change tooling system by using a 1" diameter end mill holder and a length of 1" diameter tool steel. Start by cutting the steel stock into 2" lengths. Chamfer each end with a $^1/_{16}$" radius bit. Mill a $^1/_2$" wide slot on the side of each of the cut lengths. This should resemble a Weldon shank. Next, drill and ream the 2" lengths in a high precision chuck to the sizes you will use the most. The tool bits can be held in position with a $^5/_{16}$-24 set screw that must also be drilled and tapped into the middle of the 2" length. This is an inexpensive solution to the quick-change tooling system.

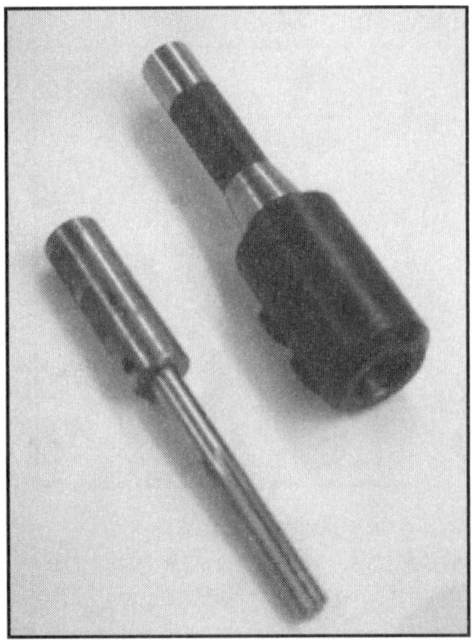

Figure 4.12 *This is a set of "home made" quick-change tool holders that are used with a 1" end-mill holder.*

Another reason for using quick-change tooling is to be able to set *tool length offsets* (TLO). By setting the tool length (how much of the tool sticks out of the holder in relationship to the top surface of the workpiece) you can use the same depth location time after time.

If you had to change collet sizes and reinstall the same set of tools (center drill, drill, tap, etc.), the tool length will be different every time. This would also change the depth of each cut when you brought the quill or table to the same depth as before. This becomes a handy convenience when you are center drilling and drilling the same size hole over and over.

Drill Chuck

Drill chucks are a definite *must-have* for any mill. They allow quick installation and removal of drill bits for drilling operations. First, realize that drill chucks will not provide the support necessary for high precision work. However, when hole locations are first center-drilled, holes can be placed with a respectable degree of accuracy. Drill chucks are made to be used with drill bits, not end mills!

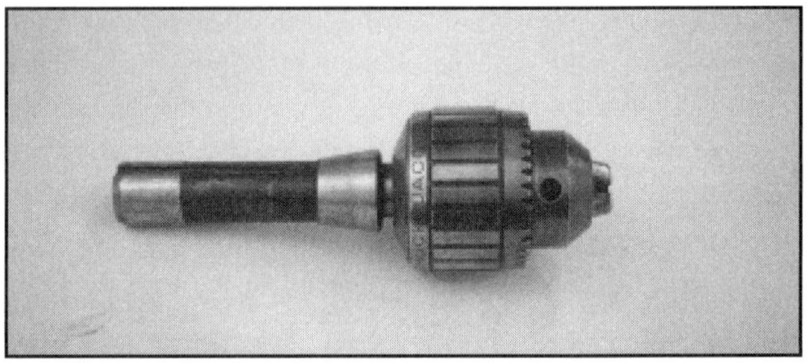

Figure 4.13 *A typical drill chuck. This one uses a key to tighten and loosen the holding jaws.*

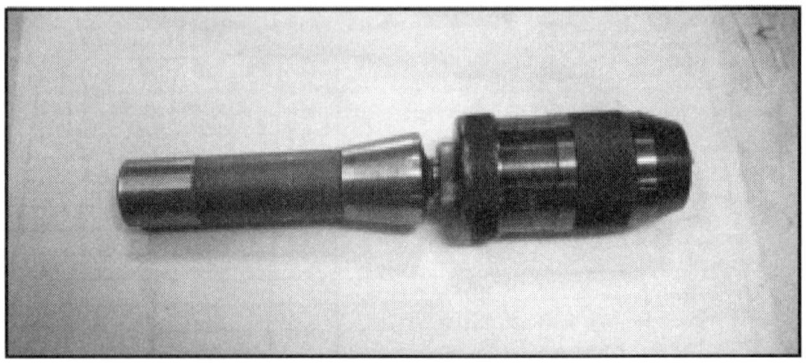

Figure 4.14 *This is a "keyless" chuck. The jaws are tightened and loosened by hand tightening.*

Drill chucks are tightened with a key, or by hand (keyless chuck), to clamp the drill. Chucks are normally mounted onto a *straight-shank* and held with a collet, or with a special *Jacobs Taper* (JT) adapter. Chuck sizes are determined by the maximum amount that the jaws will open. The majority of drill chucks used on turret mills will hold at least a $^1/_2$" diameter tool. A $^5/_8$" diameter capacity chuck would be a better purchase because often a tool just slightly larger than the $^1/_2$" limit must be used for drilling.

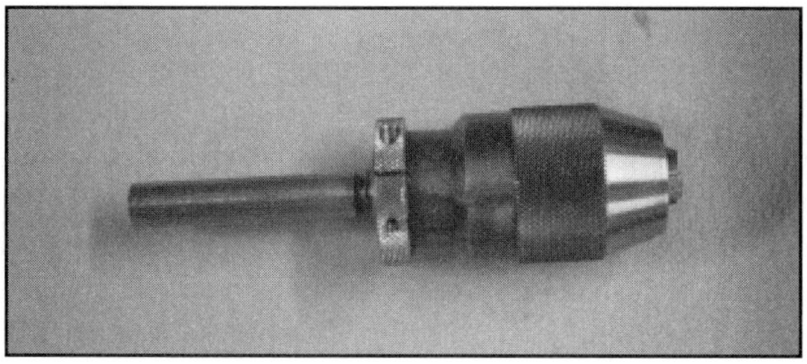

Figure 4.15 *Chucks must be mounted onto an arbor before they can be used.*

When purchasing a drill chuck, there will be questions that need to be answered before the acquisition: what will the capacity be, and what size adapter will be needed to mount the chuck? Chucks are manufactured in many sizes ranging from completely closed (0), to a maximum opening of approximately 1" (25.4 mm). The majority of drill chucks are mounted onto a tapered shaft which conforms to the Jacobs standard (JT) and are found in the following sizes: 0-JT, 1-JT, 2-JT, 3-JT, 4-JT, 5-JT, 6-JT, and the 33-JT.

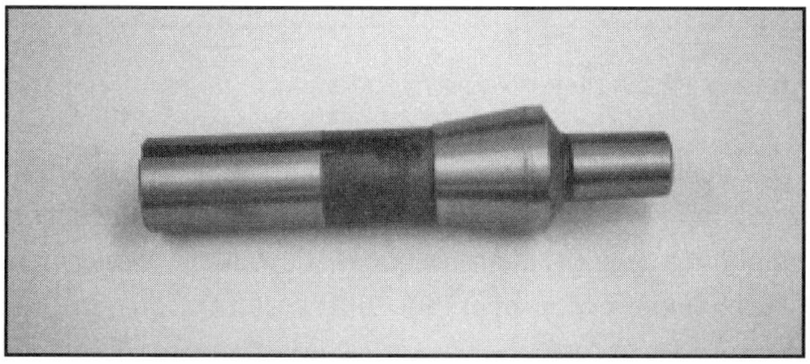

Figure 4.16 *An R-8 to Jacobs Taper arbor is used to mount a drill chuck. Drill chucks can also be fitted to straight shank adapters.*

The chuck closure mechanism will use a plain- or ball-thrust bearing. The ball bearing type is designed for production drilling operations and reduces friction in the jaw closing mechanism. As a result, the operator can apply greater gripping force onto the drill shank.

Chuck keys used to tighten the drill will be matched to the particular size chuck you use. Do not attempt to use a key from one size chuck to fit another. You will only destroy the key and/or the teeth on the chuck — either way, you lose. When finished using the drill chuck, insert the key into the jaws and tighten it hand-tight. This is so you don't have to waste time looking for it. Some machinists attach a cord to the key, this can then be tied or bolted to the machine. This is not a bad idea, unless you use several different sizes of chucks on the same machine.

Keyless chucks are nice because you don't have to worry about where you left the chuck key. By design, their gripping power increases in proportion to the torque (power) requirement of the drilling operation, which is another nice feature.

Morse Taper Adapters

Certain cutting tools found in the shop will use a *Morse taper* (MT). Drill chucks and certain large-diameter drill bits quickly come to mind. The Morse taper tool holding system has been used in shops for many years. The most common (i.e., *practical*) Morse taper sizes used on the mill will be #2 and #3. Commercially available R-8 adapters can be acquired to allow the use of Morse taper tools.

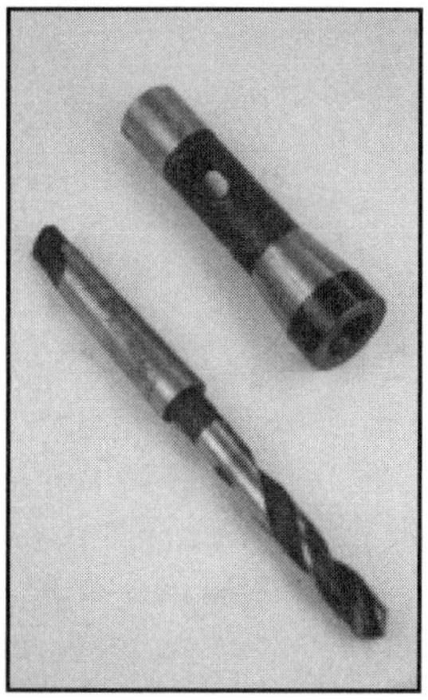

Figure 4.17 *A Morse taper adapter is necessary when using large-diameter drills with a Morse taper end.*

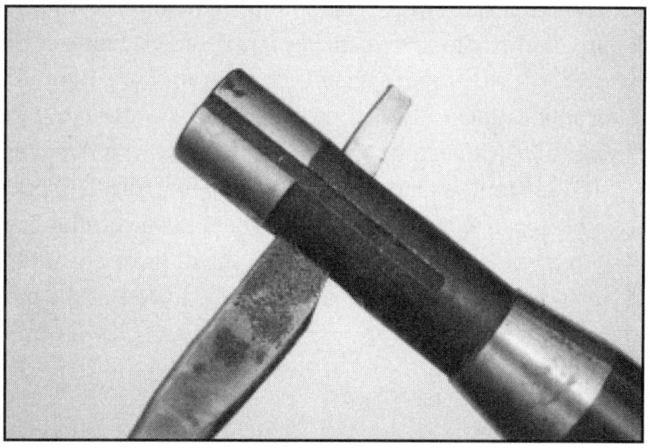

Figure 4.18 *A tapered drift is used to remove drill bits from the adapter.*

Installation of the tool into the adapter is quite easy – align the tool tang with the opening in the adapter and snap it into place. Tap the end of the adapter onto a heavy metal block to seat the tool properly. **Note**: Do not use the table of the mill for this purpose! Removal of the tool from the adapter requires a *drift*, or *center*, key. By inserting the tapered key into the long narrow opening and giving it a light *whack* with a hammer, the tool will easily separate from the adapter.

Sleeve adapters are available for mating a #1 or #2 Morse taper tool into a #3 holder, or #2 into a #3. Common Morse tapers sizes found in the machine shop include the following: #1, #2, #3, #4, #5, and #6.

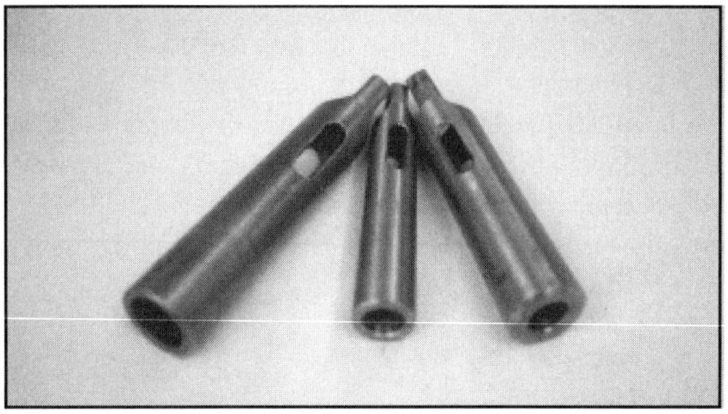

Figure 4.19 *Morse taper sleeve adapter can allow the use of smaller size drills. In our example, a #1 MT drill is held with an R-8, #3 MT adapter.*

Flycutter

A flycutter is a very handy tool that is used to cut a fairly large diameter or area. They typically use a square-tool bit to cut or bore the workpiece. Tool bits can be ground to suit a particular material or machining application. The majority of flycutters place the tool bit at an angle (around 15°) in relationship to the top of the workpiece surface. Flycutters are made in three basic configurations: the straight shank type, the R-8 shank type, and the insert type. Cutters range in diameters from $^{3}/_{4}$" (19 mm) diameter to several inches (cm). Larger-diameter cutters, over 4" (101.6 mm), should be used for very light-duty cutting applications only. The vertical knee-type turret mill was not made to machine large amounts of material with flycutters, e.g., 10" (250 mm) diameter and .125" (3.0 mm) depth of cut in steel. The machine spindle bearings will not live very long under such adverse conditions. Straight shank tools are sold in individual sizes, or in sets containing three to five of the most popular sizes.

Figure 4.20 *This flycutter has an integral R-8 shank. It can be fitted with a $^{3}/_{8}$" square high-speed or cemented-carbide tool bit.*

Cutters with the R-8 shank are typically found in the 3" (75 mm) diameter range and use a $^{3}/_{8}$" (10 mm) square tool bit. A series of set screws (typically, three) are used to clamp the tool bit into the cutter body.

For the rigors of production, insert-type flycutters use replaceable carbide inserts. This is the cutter of choice when high part counts are anticipated. Inserts may be triangular, square, round, or diamond shaped. Each insert type has a particular cutting specialty. Insert manufacturers have full-time field engineers who do nothing but make recommendations as to the tool holder type, diameter, insert shape, and number of inserts required for material applications.

Figure 4.21 *An insert-type flycutter.*

Specialty flycutters may place the cutting tool bit parallel to the surface of the workpiece in order to machine a flat surface. The cutting edge geometry (angle) is dependent upon the type of surfacing or boring operation to be performed.

Figure 4.22 *This special flycutter is used to resurface the cylinder seating area on air-cooled automotive engine cylinder heads.*

For cutting light alloy metals (aluminum, etc.), you may want to consider using a *twin-bit* flycutter. This increases the speed at which the workpiece can be machined. They are also light in weight and normally use two square tool bits. Twin-bit flycutters are available in various diameters, but 5" and 7" diameters are the most common. Some caution must be used when flycutting any material. Do not be overly aggressive with feeds, speeds, and depths of cut when using flycutters.

Figure 4.23 *A twin-bit (blade) flycutter.*

Square tool bits are ground from solid *high-speed steel* (HSS) blanks. Cutting and relief angles are carefully considered to maximize the efficiency of the cutter.

Some square shank tools are classified as *cemented carbide* tools. They have a piece of carbide brazed onto a high carbon steel shank. Various standard cutting edge configurations can be ordered for cemented carbide tool bits.

Again, the type of material to be machined dictates the cutter angle geometry. Check your machine tool handbooks for specific recommendations. Here are a couple of suggestions that will provide you with good (or better than average) surface finishes. Grind a radius on the tip of the tool rather than a sharp chisel point. The more *contact area* the cutter has with the surface, the smoother the finish will be. Use a positive rake on the tool bit instead of a flat or negative rake unless otherwise specified by the workpiece material manufacturer. Yes, material manufacturers do specify speeds and feeds, and tool geometry for their products. Give them a call.

Figure 4.24 *A left- and right-hand chisel point carbide tipped tool bit.*

Figure 4.25 *A right-hand round nose carbide tipped tool bit.*

Face/Shell Mill Arbors

When large amounts of material must be removed from a part, a *face* or *shell mill* will typically be used. They contain multiple tool bits or cutter teeth. With each bit carrying a part of the cutting load, faster speeds and higher rpm rates can be used. These mill heads are available in a variety of sizes (diameters) and number of cutter bits/teeth. Adapters used to hold face or shell mills have diameters of 1", 1-$^1/_4$", and 1-$^1/_2$". Larger diameter sizes may be available, but their use will eventually decrease the life of the spindle bearings. Face mill arbors have two *drive dogs* that must be aligned with the cutter. Never try to use an arbor with improperly sized dogs (small drive dog and large cut-out in the cutter). The resulting excessive clearance may cause the drive dogs to be sheared off during use and operation.

Figure 4.26 *A complete shell-mill assembly. This one has replaceable tool bits.*

Stub Mill Arbors

Occasionally, a *plain mill* cutter will need to be mounted on the vertical turret mill. These cutters would normally be mounted on a horizontal mill. To simplify this installation a *stub mill arbor* is used. They come in two shank size diameters, 1" and 1-$^1/_4$", and will typically have a shank length of 2-$^1/_2$" to 4".

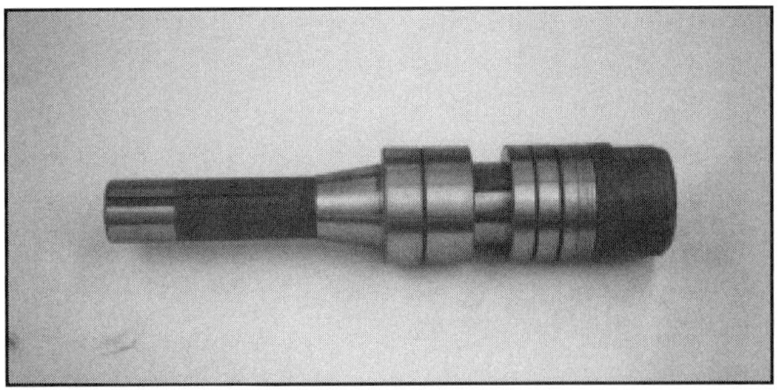

Figure 4.27 A stub arbor can be used to mount a variety of mill cutters.

Stub mill arbors can be used to hold gear cutters, single and double angle cutters, and slotting mills (all of which are traditionally performed using a horizontal mill).

Metal Slitting Saw Arbors

Metal slitting saw arbors are used to hold metal slitting and jewelers saws. They typically have a series of different sized steps onto which the saw is located. To secure the saw blade to the arbor, a cap and bolt assembly is installed. The different size steps are to accommodate the various size saws commercially available. The larger the saw blade, the larger the arbor hole.

Figure 4.28 Metal slitting saw arbors have a series of steps machined into the arbor to facilitate different hole sizes in the saw blades.

Tapping Head

When a large number of the same-size holes must be tapped, consider the use of a *tapping head*. Tapping heads can be attached directly to the spindle via a straight shank or R-8 adapter. Tapping heads dramatically increase the speed at which holes are threaded. Tapping heads do not require that the motor be reversed to remove the tap from the hole.

Figure 4.29 Tapping heads make repetitive hole tapping very easy and extremely accurate.

Selection of the tapping head will depend upon the tap size requirements. Most heads will accommodate only a specific size range of taps. Never use an undersize tapping head; this will only result in a broken tapping head.

To keep the unit from turning, tapping heads must have a torque bar. The bar keeps the head from rotating during operation. Give some thought as to where the torque bar will be positioned during the initial set-up procedure.

Figure 4.30 Tapping heads use a series of different sized collets to hold the taps.

To use the tapping head, install the unit into the spindle. Adjust the spindle speed according to the manufacturer's recommendations. **Note**: Never exceed the manufacturer's rated speed recommendations. Install the tap into the unit and install the torque bar. Upon start-up, the tap will rotate backwards – that's okay, it is normal. Once you move and engage the tap into the workpiece, the direction will immediately reverse and go forward. Use only enough pressure to keep the tap moving downward. Once you have reached the proper depth, move the spindle handle upward. At this point, the tapping head will reverse the tap and it will climb out of the hole at its own speed. Always use an approved lubricant during the tapping process.

> Do not force the tap into or out of the hole.

Boring Heads

Adjustable *boring heads* are tools that, once you have used one, you won't be able to do without it. Drill bits will perform the majority of routine drilling operations quite satisfactorily. But, for those holes with non-standard OD sizes, or for the ones that are over the maximum size of your largest drill bit, the adjustable boring head will come to your rescue.

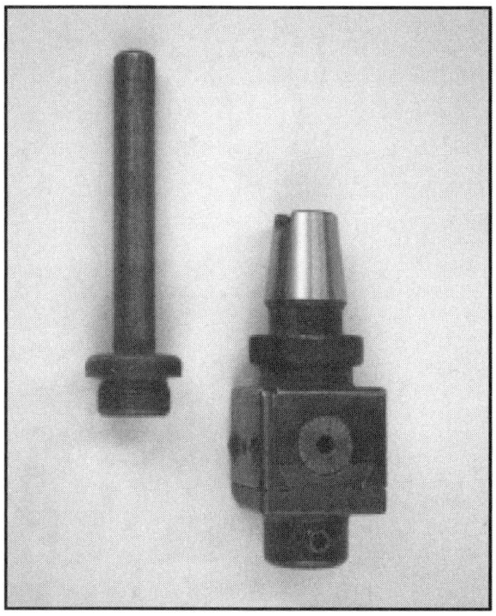

Figure 4.31 *An adjustable boring head can be used to machine just about any size hole into a workpiece.*

Because the boring head is adjustable, it will provide the greatest flexibility for size requirements. Boring heads are manufactured in various sizes ranging from 1-$^{1}/_{2}$" to 4" in diameter (or square - 1-$^{1}/_{2}$" × 1-$^{1}/_{2}$", 2" × 2", etc.). They will have a straight shank (for mounting into a collet), or they can be mounted with an R-8 adapter.

Typically, boring heads are adjustable in .001" increments, however, precision units capable of an incredible .00005" adjustment are also commercially available. These heads are unparalleled for their accuracy. One revolution of the size adjuster on a standard head will increase the bore diameter by .050". Tool bits can be installed parallel or perpendicular to the spindle axis, thus increasing the actual bore capacity of the boring head.

Take care when using this tool because boring tool bits are small in diameter. If used incorrectly, the bore will possibly wind up with a taper (larger at the top and smaller at the bottom of the bore).

Some boring heads (also commercially available) utilize a *facing* feature. These tools utilize a special internal clutch that allows a facing function. When they reach a predetermined size, or if they encounter excessive resistance to movement, the feed will cease functioning. Tool bits move outward from the center of the spindle axis. Flat surfaces or grooves can be machined in/onto a workpiece.

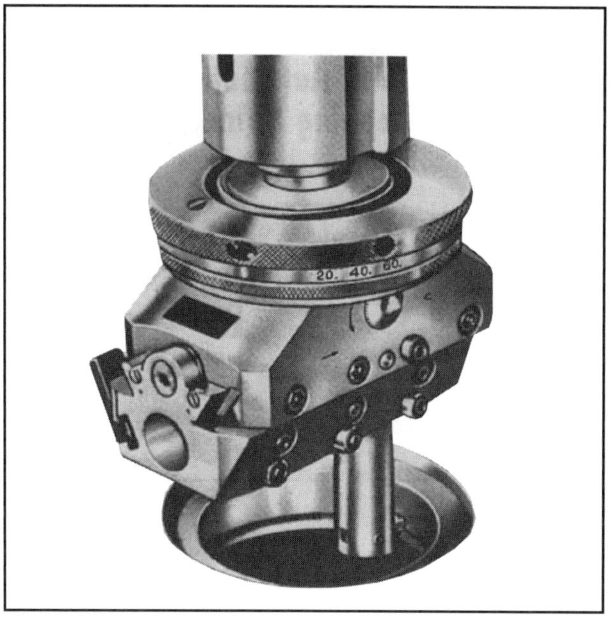

Figure 4.32 *This is an automatic facing boring head. It can be used for a variety of machining operations.*

Horizontal Milling With a Right Angle Attachment

To increase the versatility of your vertical turret mill, a *horizontal milling (right angle) attachment* can be mounted. This component will allow you to install cutters perpendicular to the table. Such would be the case if you were to machine a workpiece with a series of *stacked-up* cutters, or cutting gears. The unit fits into the R-8 spindle and uses an arbor support that is installed onto the bottom side of the V-ram. This provides stability for the horizontal cutter arbor. Arbors are typically $^{7}/_{8}$" or 1" in diameter and come in two standardized lengths of 8" and 10". The arbor support must be lubricated prior to and during use to maintain expected tool life.

Figure 4.33 The right angle attachment is very useful for operations requiring a plain-type mill cutter.

Slotting Attachment – E Head

A *slotting attachment* can be installed on the opposite end of the V-ram. It is used to cut keyways or slots into the workpiece. Installation of this tool requires the *turret* to be rotated in order for the tool to be properly positioned over the machine table. Make sure that the slotting attachment is connected to the proper voltage and is suitably grounded to the milling machine.

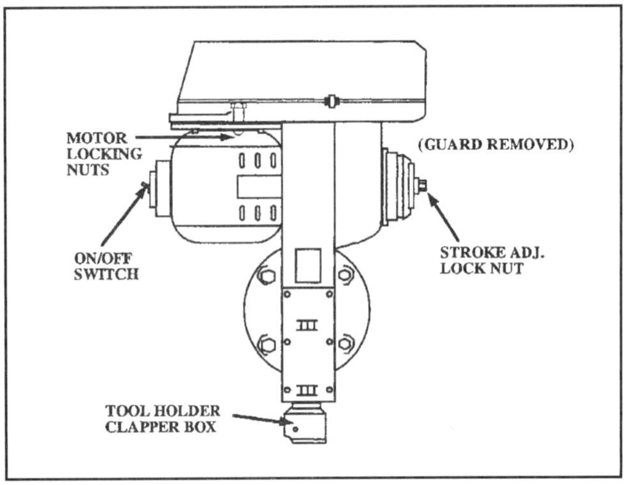

Figure 4.34 A slotting attachment. The head must be set for stroke depth and the number of strokes per minute (SPM) before it is used.

Next . . .

In the next chapter we will explore the world of cutting tools and how they are used to machine the workpiece. Proper cutter tool selection is fundamental to work quality and output. Not only will you learn the names of the various tools, but their intended applications as well. Also, take special care to learn how to calculate speeds and feeds in this chapter.

Notes:

Chapter 4 ◆ Study Questions

1. Machinist A says that the drawbar used with the R-8 collet system has 20 threads per inch, Machinist B says that the drawbar has a $^{7}/_{16}$" diameter. Who is correct?
   A. Machinist A
   B. Machinist B
   C. Both A and B
   D. Neither A nor B

2. Machinist A says that #30 tool holder system is only used for light duty machining applications, Machinist B says that 200 Series tool holders are the most common type used on the turret mills. Who is correct?
   A. Machinist A
   B. Machinist B
   C. Both A and B
   D. Neither A nor B

3. Machinist A says that R-8 collets use a set screw to secure and hold Weldon shank tool bits, Machinist B says that the Weldon shank has a small flat relief cut into the body of the cutter. Who is correct?
   A. Machinist A
   B. Machinist B
   C. Both A and B
   D. Neither A nor B

4. Machinist A says that ball bearing chucks are used for heavy duty drilling applications, Machinist B says that *keyless* drill chucks do, in fact, use a key to tighten them. Who is correct?
   A. Machinist A
   B. Machinist B
   C. Both A and B
   D. Neither A nor B

5. Machinist A says that adapters with a straight or R-8 shank are used to mount a drill chuck, Machinist B says that these adapters conform to the Jacobs Taper standards (JT). Who is correct?
   A. Machinist A
   B. Machinist B
   C. Both A and B
   D. Neither A nor B

6. Machinist A says that Morse taper drills can be installed directly into the machine spindle and held in place with the drawbar, Machinist B says that the most common Morse taper sizes used on the knee mill are the #1, #3, and #6. Who is correct?
   A. Machinist A
   B. Machinist B
   C. Both A and B
   D. Neither A nor B

7. Machinist A says that flycutters are used to cut narrow slots into a workpiece, Machinist B says that most flycutters use a square tool bit. Who is correct?
   A. Machinist A
   B. Machinist B
   C. Both A and B
   D. Neither A nor B

8. Machinist A says stub mill arbors are used to mount plain mill cutters, Machinist B says that plain mill cutters can also be mounted directly to the machine spindle. Who is correct?
   A. Machinist A
   B. Machinist B
   C. Both A and B
   D. Neither A nor B

9. Machinist A says that tapping heads rotate in *reverse* when they are initially started, Machinist B says that tapping heads must be selected according to the size of taps you will be using. Who is correct?
   A. Machinist A
   B. Machinist B
   C. Both A and B
   D. Neither A nor B

10. Machinist A says that horizontal milling attachments extend the usefulness of the turret mill, Machinist B says that the E Head is used for cutting keyways and slots. Who is correct?
    A. Machinist A
    B. Machinist B
    C. Both A and B
    D. Neither A nor B

Chapter 5
Cutting Tools

Primary to all cutting tool performance is life expectancy. All tooling must provide a certain level of performance that is acceptable to both the machinist and end user of the product. Machinists must perform the work quickly to maintain cost control for machining time and tooling costs. The end user must approve all costs and quality issues, which conform to his or her product application.

Cutting tools have a limited life span. This means that replacement and/or resharpening must be performed periodically. Tool life is varied, as is the type of materials and machinists who cut them. Yet, there must be a starting point from which to determine the speed and feed rate of a tool. Factors taken into consideration include: hardness of the material to be cut, the depth of cut, chip load, abrasiveness of the material, and the amount of time it will take to perform the machining process.

Figure 5.1 *Tools have a limited life span. Once cutting edges have disappeared, it's time to replace or resharpen the cutter. Note the darkened (color is dark blue) end. This represents excessive heat build-up in the tool bit which was ruined due to excessive cutter (rpm) speed.*

All materials have recommended cutting speeds, as can be found in machinists' reference materials, or published by the material manufacturer. All references presented here will suit the majority of *general* applications. Production shops, however, will want to know if they can increase the speeds and feeds of the tooling. Through experimentation you will discover just how much a tool can be *pushed*. Increasing speeds and feeds will ultimately shorten the life span of any tool. Running a tool too slowly will increase the machining time and increase heat build-up in the tool. Too fast, and the tool does not have sufficient time to remove material in a satisfactory manner and will leave behind a rough surface finish. There is a delicate balance between the speeds and feeds and the cost of doing business. Increasing speed will increase tooling costs and require the machinist to stop and change tooling more often.

In this section we want to examine the various types of cutting tools typically associated with milling projects. We want to investigate what the tools are made of, and identify the differences between drill bits, end mills, and specialty cutters commonly used for milling operations.

Figure 5.2 *A selection of cutting tools used on a knee-type mill.*

Cutting tools are made of metal. Logical thinking should tell us that cutting tools should be harder than the material they are to cut. Soft workpieces can be cut with standard (less costly) tooling. Exotic and space age materials may often require cutting tools and holders more costly than the job itself.

The majority of cutting tools are made from special steel alloys or hybrid compositions. Among the most common tools used for milling operations are *high-speed steel* (HSS) and *carbide*. High-speed tool steels are an alloy that contain large amounts of molybdenum (Mo), chromium (Cr), and vanadium (V), combined together with the basic fortifier, carbon (C). High-speed steels are divided into two groupings: tungsten and molybdenum. Tungsten tools will have a prefix *T* before the number, which designates its grade. Molybdenum alloyed tools will have an *M* prefix.

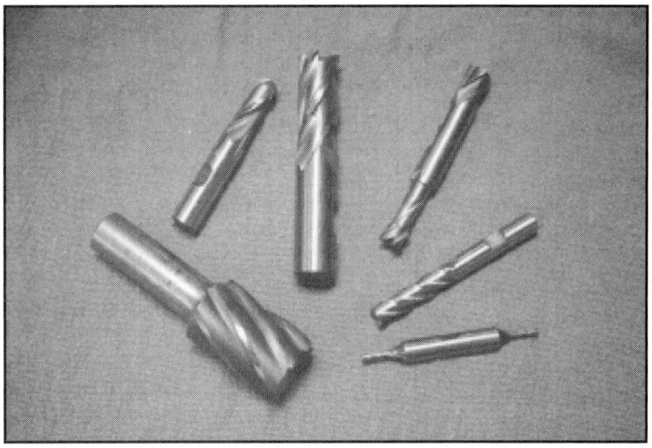

Figure 5.3 *This selection of high-speed steel (HSS) tools includes standard HSS, and cobalt types.*

High-speed steels containing tungsten have been almost entirely replaced by the molybdenum (moly) types. Most common of these moly steels are M2 and M40. Hardness of these steels will range from 63 R_c to 65 R_c. M40 steels will approach a hardness of 70 R_c. In general, high-speed steels will retain their effective hardness at temperatures ranging between 1000° and 1100° F, high enough to maintain sufficient hardness during most milling operations.

To increase the brittleness of high-speed steel, cobalt can be added to the alloy mixture. Amounts are specified as 5%, 10%, or 15% cobalt. Cobalt tools are used primarily for roughing operations where high chip loads and fast feed rates are the norm. All tools made from HSS are ground from solid blanks or bars.

High-speed steel is the industry standard by which all other cutting materials are compared. They can be resharpened numerous times if used properly.

Rockwell C Hardness (R_c)		Equation for converting R_C to Brinell Hardness (BHN)
from	to	
21	30	BHN = 5.97 × HRC + 104.7
31	40	BHN = 8.57 × HRC + 27.6
41	50	BHN = 11.158 × HRC + 79.6
51	60	BHN = 17.515 × HRC - 401
Rockwell B Hardness (R_b)		**Equation for converting R_B to Brinell Hardness (BHN)**
from	to	
55	69	BHN = 1.646 × HRB + 8.7
70	79	BHN = 2.394 × HRB - 42.7
80	89	BHN = 3.297 × HRB - 114
90	100	BHN = 5.582 × HRB - 319

Figure 5.4 *Rockwell C (R_C), Rockwell B (R_B), and Brinell (BHN) numbers refer to hardness. The above chart provides equations for converting Rockwell to Brinell numbers. (Courtesy of Kennametal.)*

Carbide is a sintered metal product. It is made from highly refined tungsten ores approximately 94% to 97% pure. They retain their hardness at temperatures approaching 1400° F, and higher. Carbide has a hardness of 90.5 R_c to 93.5 R_c, making it a very desirable cutting tool material.

Carbide cutting tools are ground from solid blocks, or they have cutting edges made of carbide. The big plus for carbide tooling is that it can be run up to four times faster than HSS. It is also a prerequisite for cutting certain space-age alloys. Solid tools can be, like HSS, resharpened if they are treated carefully. Carbide is less tolerant of shock loading than HSS, meaning that it cannot be dropped or banged around without chipping or breakage. Carbide is easy to distinguish from HSS: it does not have a bright shiny appearance, and it is quite heavy by comparison, with a dull gray appearance.

Occasionally, you will purchase *carbide-tipped* tooling. This makes good sense because the body can be made of a chrome-moly steel which provides good support and resistance to shock loading, while having the cutting durability made famous by carbide. These tools too can be resharpened or reset with new blades for their eventual reuse.

Figure 5.5 *This is an example of a carbide-tipped tool bit.*

Production shops will almost always use tooling with carbide inserts. Inserts can be rotated to expose a new cutting surface in the exact same location at the one before. Because the inserts are small, their replacement cost is considerably less than carbide-tipped tooling.

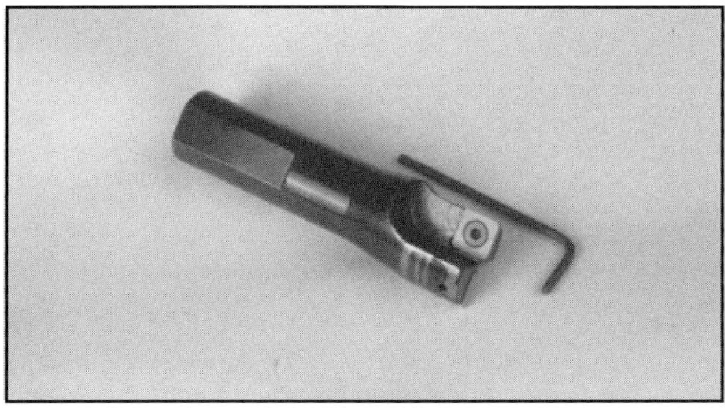

Figure 5.6 *This carbide insert is typical of cutters used in a production machine shop.*

Carbide inserts are classified as to their cutting ability. The majority, however, fall within five general industry grades: C-1, C-2, C-5, C-6, and C-7. In general, purchase the better grades of carbide when parts production demands are high. The difference in price is more than off set when you can produce 50 parts, instead of 5 or 6, with the same tool.

Industry Class	Application	Materials	Working Conditions
C-2	Turning, Boring, Milling	Cast iron, copper, brass, non-ferrous, high-temperature alloys, stone, plastics	General purpose grade for high toughness, resistance against wear at low to medium cutting speeds
C-5	Turning, Boring, Milling	Steel, cast steel, malleable cast iron, 400/500 stainless steels	General purpose with wide range of applications, low to medium cutting speeds, high feeds and depths of cut
C-2, C-7	Turning, Boring, Milling, Threading and Grooving	Cast iron, stainless steel, Ph stainless steel	TiN coated grade, with broad range of applications, crater and impact resistance at low to high speeds
C-1, C-2 C-5, C-6	Milling, Turning, Grooving Threading	Cast iron, steels, high temperature alloys, 300/400 stainless steels, Ph stainless steel	TiN coated heavy duty grade used in severe roughing and interrupted cuts at slow to moderate speeds

Figure 5.7 *Carbide classifications.*

Tool Coatings

To increase cutter tool life, more and more tools are being coated or treated with special surface coatings, most notably, *titanium nitride* (TiN) and *titanium carbonitride* (TiCN). Surface coatings are applied through a vapor deposition process. Coatings do increase the initial cost of the tooling, but the payback in increased productivity more than offsets their additional expense. TiN coatings are golden in color, TiCN coatings are dull gray in appearance.

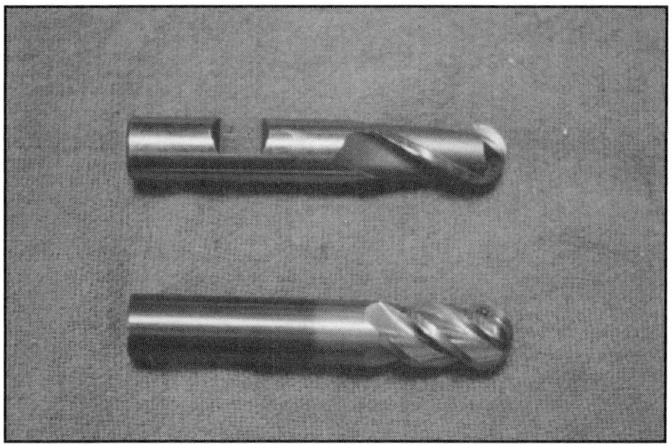

Figure 5.8 *TiN tool coatings are typified by their gold looking color. TiCN tools are typically a dark gray color.*

Care of Cutting Tools

Care of your cutting tools is as important as their selection and purchase. Tool bits should not be thrown into a drawer after you are finished using them, nor should they be stored where they will become rusty. The sharp edges of the cutting tools must be protected so that they do not become nicked or dinged. Small craters in the cutting edge will cause the tool to run hot and/or leave behind a terrible surface finish. You may think "what's the big deal, I have two flutes." This may be true, but the good flute must perform double-duty and makeup for the damaged area.

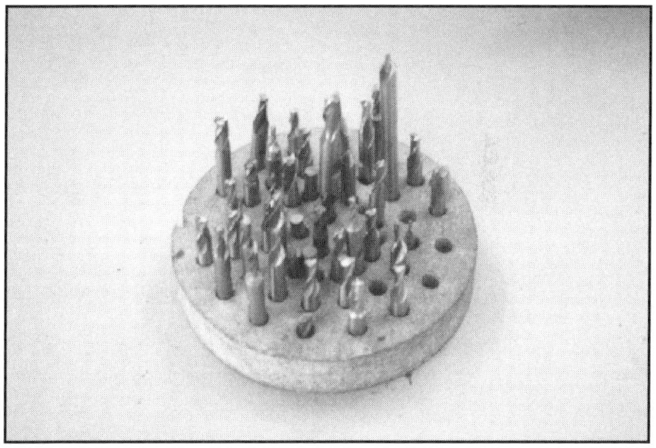

Figure 5.9 *This is an example of a tool caddy used to hold end mills. This one was made using a thick scrap of particle board.*

Store your end mills in racks designed for this purpose. Racks can be made by drilling appropriate sized holes in a wooden block. This not only allows the tools to keep a little *breathing room* between each one, but it will allow you to organize your tools by size or type. If tool bits must be stored in a cabinet or steel drawer, wrap a piece of masking tape around each bit or place a plastic sleeve over the cutter flutes.

A good practice to observe when using cutting tools is to

color the tool after it has been used to distinguish it from new tools. This can be done with a permanent marker. Also, you may wish to color-code your cutters as to the type of material you will be cutting. Cutters used to cut mild-steel may not cut a piece of bronze very well having cut the steel first. If at all possible, have your cutters sharpened for the type of metal you will be machining. The relief angles are quite different for each type.

Cutting Tools

Now that we have a good idea what cutting tools are made from, we can take a closer look at the differences between them. Learning how to specify cutting tools is *part learned* and *part experience*: sort of *been there– done that.* Cutting tools used on a milling machine continuously enter and exit the workpiece, this makes their life much more troublesome than, for instance, a lathe.

The chips removed from the workpiece are generally tapered. As the tool starts its cut the chip is very thin, until the cutter comes to fully bear upon the part. At this point the chip will have reached its maximum thickness. We call this *chip load*, and there will be more on chip loads later in this chapter.

Figure 5.10 *Chip load is very important. If too little material is removed, the job will take longer than necessary. Too much, and tool life will be shortened.*

Removing large amounts of material will require a large *flute*, or recess, for the chips to remove themselves. *Single-* and *double-fluted* cutting tools have large flutes and are typically used for *roughing* work. Roughing removes large amounts of material very quickly, but does not leave a desirable surface finish. The use of roughing tools will increase the speed by which the part can be produced on some jobs. After roughing the part to *near net size*, a *finish cutter* is installed for final stock removal and to achieve a prescribed surface finish specified on the blue print.

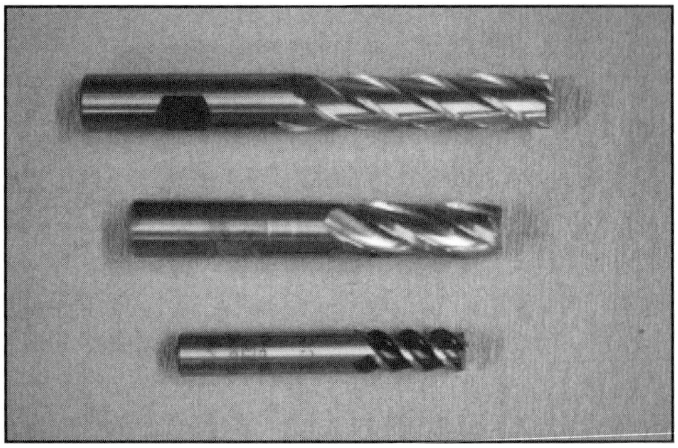

Figure 5.11 *These four-flute end mills are examples of those types used to remove stock from a workpiece.*

Conventional and Climb Milling

An understanding of the *direction of cut* is elementary to your understanding of how the cutters will remove material from the workpiece. Tools rotate in either *right-hand* (RH), or *left-hand* (LH) rotations, depending upon which direction the flutes are ground. Right-hand tools rotate in a clockwise (CW) rotation. Left-hand tools rotate counterclockwise (CCW), sometimes called *anti-clockwise*. The direction of cut is another important aspect of the cutting process — feed the cutter one way and the surface finish is poor, feed it the other, and the surface finish is very smooth.

The direction of cut is classified as *conventional milling* or *climb milling*. Conventional milling feeds the cutter against the cutter rotation. In climb milling, the workpiece is fed in the same direction as the cutter rotation. This causes the cutter to try and climb over the top of the workpiece.

Conventional milling is typically used for *roughing* the part to size because the cutter does not want to climb over the workpiece, but rather scoops the material from the bottom upward. This also allows for heavier (increased chip load) cuts. Conventional milling is also used when *pocket milling*.

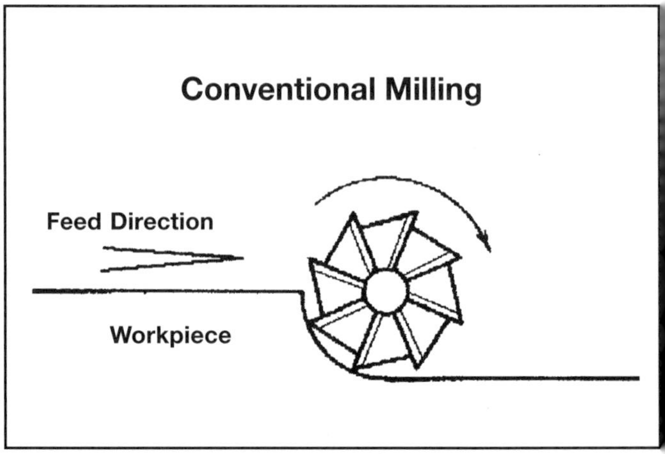

Figure 5.12 *Conventional milling.*

Climb milling is used for finish cutting after a workpiece has been roughed to size. Spindle speed should be increased and the chip load and feed rate slowed down. This allows the cutter to spend more cutting time in the same area. Remember, as the cutter is turning, the workpiece is also moving. Rapid feed rates will leave behind a rough surface finish.

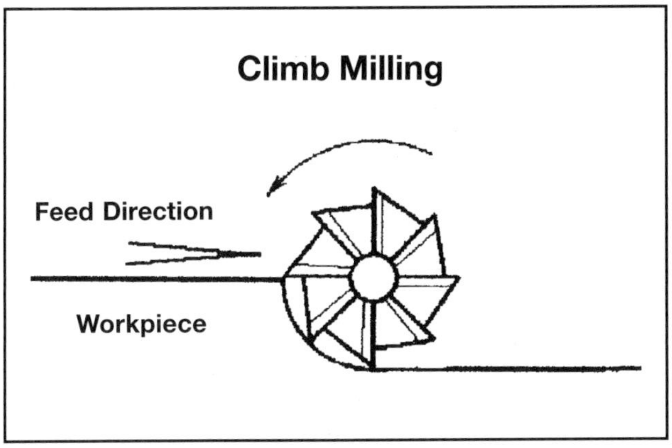

Figure 5.13 Climb milling.

Speeds and Feeds

The importance of calculating initial spindle speeds can never be stressed enough. We use the term *initial speeds* because as you have already learned, various material types will require different speeds. Now you will get a glimpse of what all this fuss is about. It would take the contents of another book to list all of the *surface speeds* required for all of the materials commercially available.

When material is to be removed from a workpiece, you will need to calculate the surface speed. Imagine, if you will, a piece of round stock with a 100-foot-long piece of string attached to it. If we spin the stock fast enough to roll up the entire length of the string in one minute, the surface speed would be 100 *surface feet per minute* (sfm). When we remove stock from a workpiece, we are essentially removing the string. The string, of course, is the length of the chip(s) we remove. But how fast do you need to spin the tool in order to remove the stock in one minute? Easy, with a simple calculation.

$$\frac{\text{surface feet} \times 4}{\text{part or cutter diameter}} = \text{spindle rpm}$$

Here's an example:

$$\frac{120\ (\text{sfm}) \times 4}{.750} = \frac{480}{.750} = 640\ \text{rpm}$$

From our example, you can see that a .750" diameter cutter would have to run at 640 rpm to achieve a surface speed of 120 sfm. This formula is used for all cutting tool applications. Make sure that you learn it now before you move on in this chapter.

Still seem a little confusing? That's okay, here's another formula that will get you the same speed information, only this time it will work in *inches per minute* (ipm). Your first question will be, "But what's the difference between *sfm* and *ipm*"? The answer is... enough so that you can burn up your cutting tools or the workpiece in pretty short order. Where sfm is a measure of the continuous chip length, ipm is the distance the tool or workpiece moves from one point to another (*a linear movement*) in the space of one minute; sfm is a *rotary* measurement.

Measurement of ipm is pretty simple. Place a non-permanent mark with a china marker onto the table and knee cross slide. Start the machine table feed and observe your watch or clock to determine how far the table travels in one minute. Adjust the feed dial on the power feed *faster* or *slower* to achieve the desired feed rate. Some milling machines may have a mechanical feed drive already calibrated in ipm. If your machine is so fitted, simply adjust it to the desired feed rate.

$$\frac{1000 \times \text{feed rate (inches per minute)}}{\pi \times \text{diameter}} = \text{spindle rpm}$$

Here's another example:

$$\frac{1000 \times 1^1/_2}{3.1416 \times .750} = \frac{1500}{2.3562} = 636\ \text{rpm}$$

As you can clearly see, both solutions will get you to the same *general* speed 636–640 rpm. When dealing with speed and feed issues, be sure to remember that all calculations are made using high-speed steel (HSS) as a benchmark figure. The numbers, if computed correctly, will get you in the ballpark as far as being able to produce parts without wiping out the tool or part. Many inexperienced machinists will typically tend to crankup the speed and start cutting. To their uneducated surprise, the tooling fails very quickly. Simply put, calculate the correct speeds and feeds for each part you produce. The dividend will be money in your pocket.

Can speed and feed be increased over what the calculations dictate? Yes. But, be very conservative in your upward adjustments. Remember, you are dealing with high-speed steels most of the time. When different tooling (cobalt, tungsten, or carbide types) is used, consult the tooling manufacturer for feed and speed recommendations. As a general *rule of thumb*, carbide can be run at 4 times the speed of HSS. *Example:* 300 rpm for a high-speed steel cutter would translate to 1200 rpm for a carbide tool.

Chip Loads

Chip loads, like speeds and feeds, are given as a general specification for a specific material type. Each cutter tooth should carry a fair share of the load to remove material more quickly than a single blade or tooth. Additional blades can remove significantly more material than a single blade or tooth.

One of the *sins* most often committed is *under-feeding* the tool into the part. This occurs when the cutter does not cut enough stock during operation.

Cutting tools must be sharp, and ground to cut specific material types. Dull tools can and will not cut efficiently. They will, however, produce large amounts of heat because the teeth are rubbing against the part, not cutting it.

As a general rule, the following recommendations are for common material types:

Material Type	Cutter Diameter	Chip Load Per Tooth
Aluminum	1/4"	.0020"
"	3/8"	.0030"
"	1/2"	.0050"
"	3/4"	.0060"
"	1"	.0070"
"	2"	.0090"
Brass	1/4"	.0020"
"	3/8"	.0030"
"	1/2"	.0020"
"	3/4"	.0040"
"	1"	.0050"
"	2"	.0060"
Bronze	1/4"	.0010"
"	3/8"	.0020"
"	1/2"	.0030"
"	3/4"	.0030"
"	1"	.0040"
"	2"	.0050"
Cast Iron	1/4"	.0010"
"	3/8"	.0020"
"	1/2"	.0025"
"	3/4"	.0030"
"	1"	.0035"
"	2"	.0050"
Low Carbon Steel	1/4"	.0010"
	3/8"	.0020"
	1/2"	.0020"
	3/4"	.0040"
	1"	.0050"
	2"	.0070"
High Carbon Steel	1/4"	.0010"
	3/8"	.0020"
	1/2"	.0020"
	3/4"	.0030"
	1"	.0030"
	2"	.0040"
Medium Alloy Steel	1/4"	.0005"
	3/8"	.0010"
	1/2"	.0010"
	3/4"	.0020"
	1"	.0030"
	2"	.0030"
Stainless Steel	1/4"	.0010"
	3/8"	.0020"
	1/2"	.0020"
	3/4"	.0030"
	1"	.0040"
	2"	.0050"

Figure 5.14 *Typical chip loads.*

Depth of Cut

Cutter *depth of cut* is limited by: the amount of material which must be removed, the amount of power available to turn the machine spindle, how the part is held, tool diameter, and the number of flutes. The rule of thumb in this case is: cutter depth should be from half to one and one half ($^1/_2$ to $1^1/_2$) the diameter of the cutter in most iron based materials when using standard high-speed steel or solid carbide end mills. Roughing end mills, designed to *hog out* (rapidly remove) the stock very quickly, may run at a depth equal to one to two times (1–2) the tool diameter.

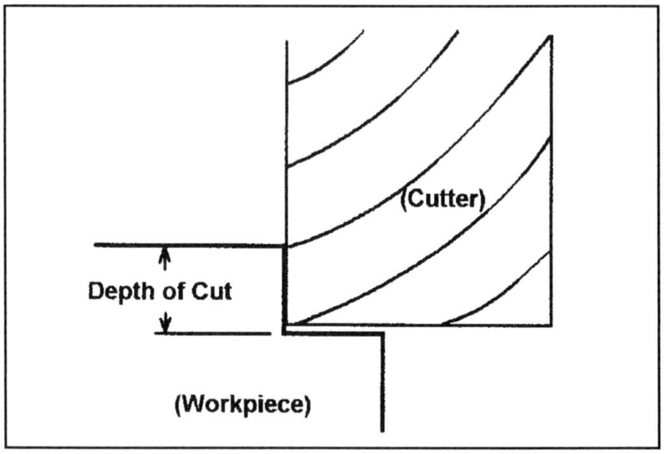

Figure 5.15 *Depth of cut is the distance that the cutter is plunged into the workpiece.*

Can end mills be used to cut even greater depths? The answer is dependent upon the chip load, the length of the cutter, the rigidity of the clamped workpiece, material type, and machine horsepower. Common sense should prevail at this point. A long or extended length small diameter end mill cannot be expected to cut at full cutter depth through a solid piece of steel. The flute would deflect severely from a lack of support and cause the tool to either break, or not cut a true 90° edge. Insert- type cutter tools offer better performance when machining deep cuts, but are not typically made in small diameters.

With practice and careful observation you will find that deeper depth of cuts are certainly possible. However, start your additional depth of cuts in small increments. Some machinists call this *sneaking up on it*. Listen to the machine, it will tell you when enough is enough. Don't be overanxious and *run up* on the part; the result may be broken tools and personal injury.

Drill Bits

Drill bits are some of the most universal cutting tools used in any shop. They remove material with two cutting edges at the same time while engaged into the work. Drills come in a variety of sizes, styles, lengths, and material types. Sizes include: SAE, or fraction sizes ($^{1}/_{64}$" –1"); number (1–80); letter (A–Z); and ISO or metric sizes (0.01–25 mm). One may ask why we need so many different sizes, it's because there are so may diverse drilling needs – e.g., tap drills, metering orifices, press-fits, etc. For a comprehensive list of drill sizes and decimal equivalents, see the *Appendix* at the back of the book.

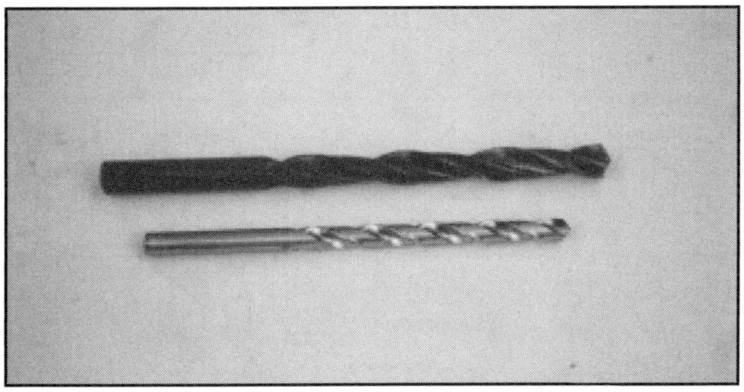

Figure 5.16 *Typical jobbers length drill bits.*

It's important to keep your drill bit sharp. Dull or incorrectly sharpened bits will not make *on-size* or *straight* holes. The majority of drill bits supplied to machine shops are called *jobbers length* drill bits. They are what you would purchase for the majority of all drilling operations in a machine shop.

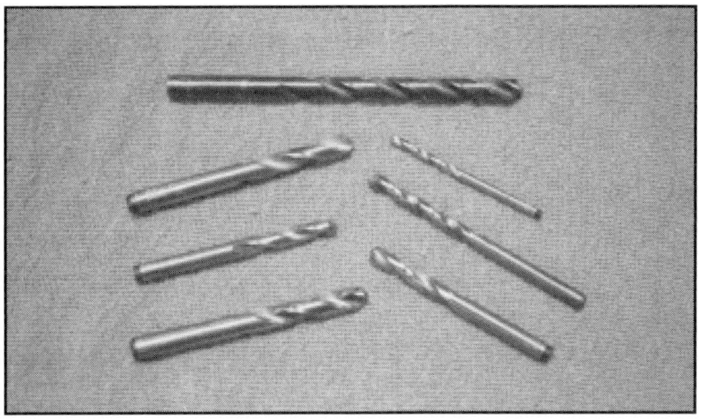

Figure 5.17 *Screw machine length drill bits; the jobbers length drill on top is for comparison purposes only.*

There are also short drill bits called *screw machine length*. They are used primarily on screw machines. These short drill bits also find many applications for milling operations. Because of their length, they can reduce spindle travel and thereby increase production.

For deep-hole drilling there are *aircraft* drills. These drills were designed for the aircraft industry, primarily for drilling rivet holes in the airframe. They have a long shank and a relatively short flute area. They are available in 12" and 18" lengths. **Note:** You should center drill the hole location before you use an aircraft drill because of the bit's shank length.

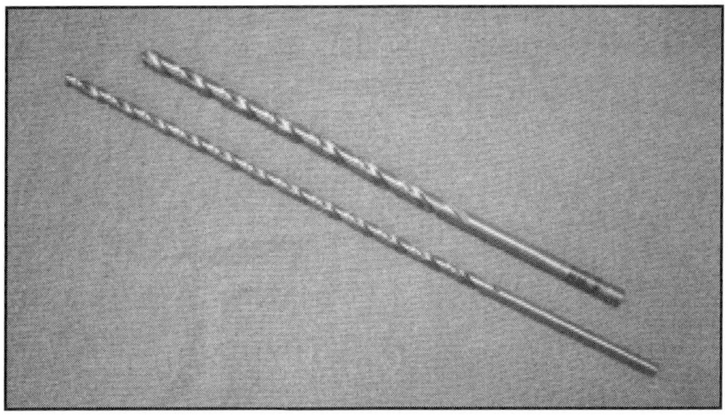

Figure 5.18 *An aircraft drill bit is considerably longer than a jobbers length drill.*

To remove larger amounts of material you may find that *reduced diameter* drill bits may fit the bill. This group of drills is often referred to as *Silver & Deming* drills. They allow you to drill holes over $^{1}/_{2}$" in diameter and still use a $^{1}/_{2}$" collet. Because the shanks are turned down to $^{1}/_{2}$" diameter, they may also reduce tooling costs in some cases. They are available in sizes up to 1" diameter.

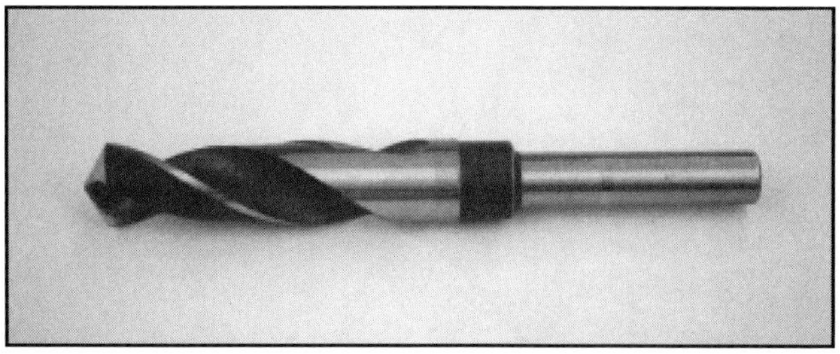

Figure 5.19 This is an example of a reduced shank (Silver & Deming drill).

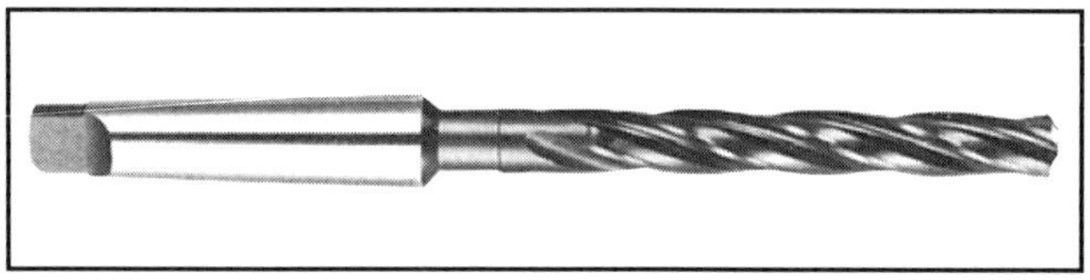

Figure 5.20 A high speed steel tapered shank drill.

To make the task of keeping drill bits separated, always keep them in a *drill index*. An index arranges the bits in ascending size order starting from the smallest to the largest. Purchasing drill bits in sets will generally include a drill index as part of the purchase. Drill sets include: fraction, number, letter, metric, and blank sizes. A drill blank is a drill bit before it is machined.

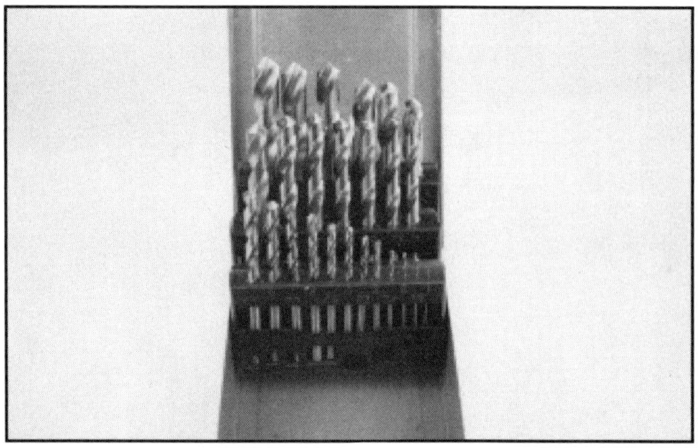

Figure 5.21 A set of metric drills in an index.

Center Drills

In days of old, a blacksmith would use a center punch to mark the location where a drilled hole was to be located. Unfortunately, the center punch is only accurate to within $^{1}/_{32}$" (0.75 mm); a better solution is necessary. *Center drills* are used to precisely locate a hole into a workpiece. They have a large shank diameter as compared to the actual (pilot-)drill end. This increase in diameter provides the center drill with much needed rigidity. Center drills come in several sizes and lengths. Sizes typically range from #00000 to #10. Standard lengths range from $1^{1}/_{4}$" (#00000) to $3^{3}/_{4}$" (#10). For sizes #1 to #8, extra long center drills are found in 3", 4", 5", and 6" lengths.

The majority of commercially available center drills have a 60° straight taper. This is the taper found on most lathe centers. There is also a *radius* type, which has no pilot-drill end but cuts a continuous radius and is also compatible with a 60° lathe center. These types of center drills allow for increased spindle speeds and reduce the possibility of pilot-drill breakage in the hole.

Center Drill Dimensions - Standard Length

Size	Body Diameter	Point Diameter	Point Length	OAL
00000	$^{1}/_{8}$"	.010"	.010"	$1^{1}/_{4}$"
0000	$^{1}/_{8}$"	.015"	.015"	$1^{1}/_{4}$"
000	$^{1}/_{8}$"	.020"	$^{3}/_{64}$"	$1^{1}/_{4}$"
00	$^{1}/_{8}$"	.025"	$^{1}/_{16}$"	$1^{1}/_{4}$"
0	$^{1}/_{8}$"	$^{1}/_{32}$"	$^{1}/_{16}$"	$1^{1}/_{4}$"
1	$^{1}/_{8}$"	$^{3}/_{64}$"	$^{3}/_{64}$"	$1^{1}/_{4}$"
2	$^{3}/_{16}$"	$^{5}/_{64}$"	$^{5}/_{64}$"	$1^{7}/_{8}$"
3	$^{1}/_{4}$"	$^{7}/_{64}$"	$^{7}/_{64}$"	2"
4	$^{5}/_{16}$"	$^{1}/_{8}$"	$^{1}/_{8}$"	$2^{1}/_{8}$"
4-$^{1}/_{2}$	$^{3}/_{8}$"	$^{9}/_{64}$"	$^{9}/_{64}$"	$2^{1}/_{2}$"
5	$^{7}/_{16}$"	$^{3}/_{16}$"	$^{3}/_{16}$"	$2^{3}/_{4}$"
6	$^{1}/_{2}$"	$^{7}/_{32}$"	$^{7}/_{32}$"	3"
7	$^{5}/_{8}$"	$^{1}/_{4}$"	$^{1}/_{4}$"	$3^{1}/_{4}$"
8	$^{3}/_{4}$"	$^{5}/_{16}$"	$^{5}/_{16}$"	$3^{1}/_{2}$"
9	$^{7}/_{8}$"	$^{11}/_{32}$"	$^{11}/_{32}$"	$3^{5}/_{8}$"
10	1"	$^{3}/_{8}$"	$^{3}/_{8}$"	$3^{3}/_{4}$"

Figure 5.22 60° center drills come in a variety of sizes and lengths.

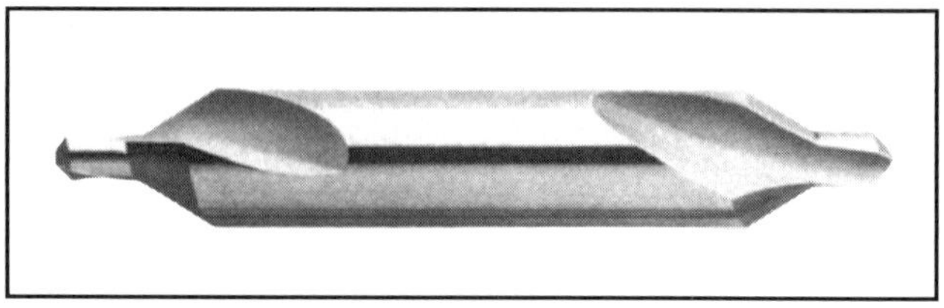

Figure 5.23 A typical 60° center drill.

The use of center drills will also allow you to start a pilot-hole into a workpiece that does not have the work surface parallel to the machine table, as would be the case with some castings. For those situations where a hole must be drilled into a steeply tapered workpiece, you may have to use a center-cutting end mill to drill (a loose term here) a flat spot for the center drill to start its hole.

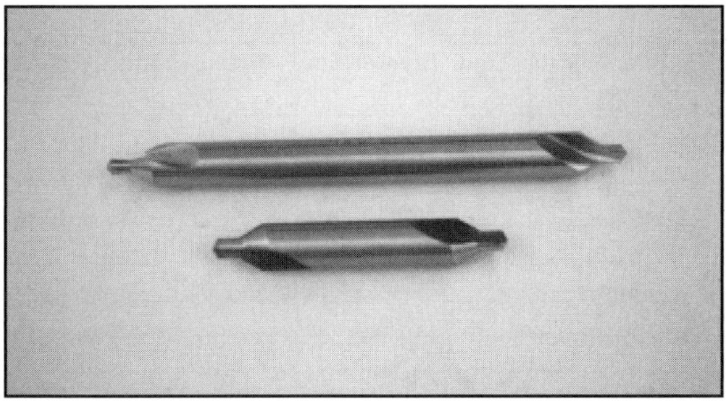

Figure 5.24 *In comparison, a long center drill is much longer than a short one. Longer center drills reduce the amount of spindle travel and should be considered for production purposes.*

The majority of machine drawings used to produce parts will specify a 90° chamfer on drilled holes. The chamfer is used to eliminate sharp edges from the part and aid in eventual assembly. To that end, center drills with a 90° taper should be used to start the hole location. This also means that you will have to use a drill with a larger diameter than the bit that you will use to drill the hole. Depth control will also be another issue to deal with. Most chamfer specifications will include a width spec. Too little (or too much) depth, and the chamfer is wrong. Check your specifications and make sure that it is correct. By using the 90° center drill you will eliminate an extra machining step, which will mean time saved.

End Mills

The most common cutter used to remove material with a milling machine is the *end mill*. They come in a wide variety of sizes and flute designs. The majority of end mills are made from high-speed steel. *Cemented* (carbide-tipped) and solid carbide end mills are also commercially available.

When large amounts of material must be removed from the workpiece, take more than one cut. How many cuts? That depends upon the factors previously discussed. It is considered good form to *hog* the majority of the material out with roughing end mills until the part is within .010" to .020" (0.25 to 0.50) of finish size, then remove the remaining stock with a finish cutter running at a slightly higher spindle speed and climb cutting.

End mills are made with same-size, oversize, and reduced shank diameters. The shank is that part of the end mill that is held in the collet or tool holder, e.g., $^{3}/_{8}$ diameter cutter with $^{3}/_{8}$" diameter shank (same-size), $^{1}/_{4}$" diameter cutter with $^{3}/_{8}$" shank (oversize), and $^{3}/_{4}$" diameter cutter with $^{1}/_{2}$" shank (undersize).

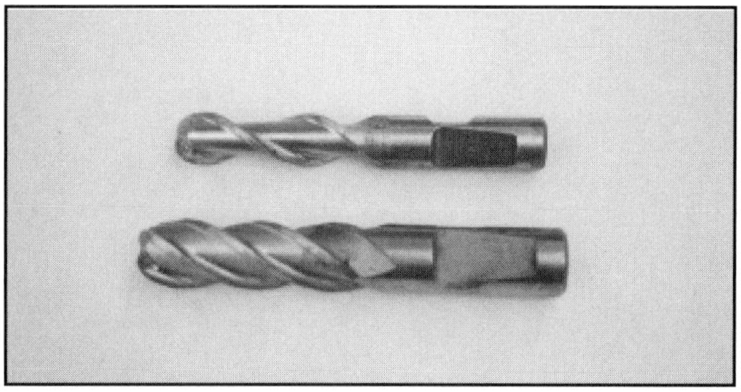

Figure 5.25 *Same-size shank end mills.*

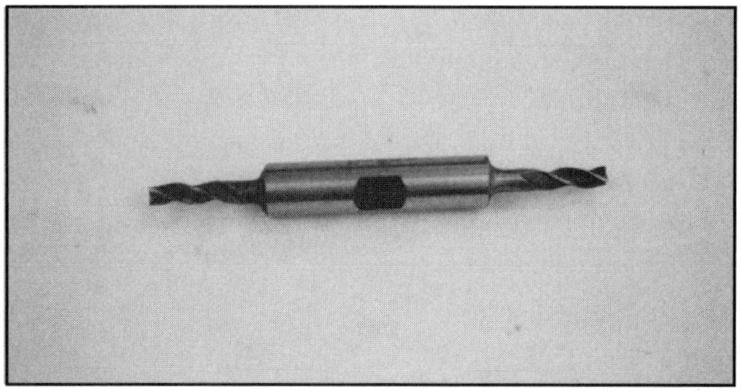

Figure 5.26 *An oversize shank end mill.*

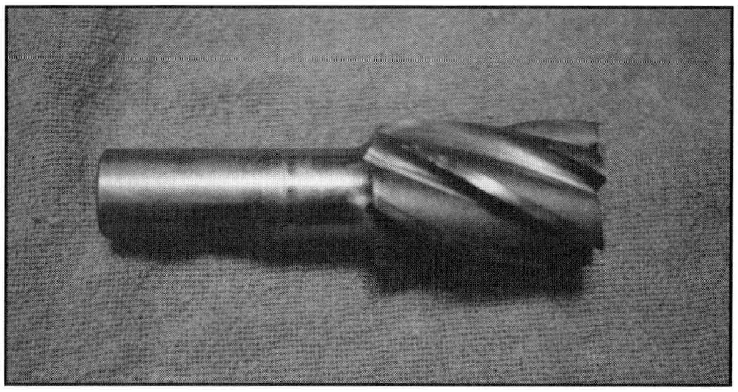

Figure 5.27 *An undersize shank end mill.*

When purchasing an end mill you will have to specify either single- or double-ends. A single-ended end mill will have only one end of the tool ground for cutting. A double-ended tool has two cutting ends. When this tool wears one side out, it can be turned end for end to expose a new cutter. Double-ended tools are slightly higher in cost than single-ended types.

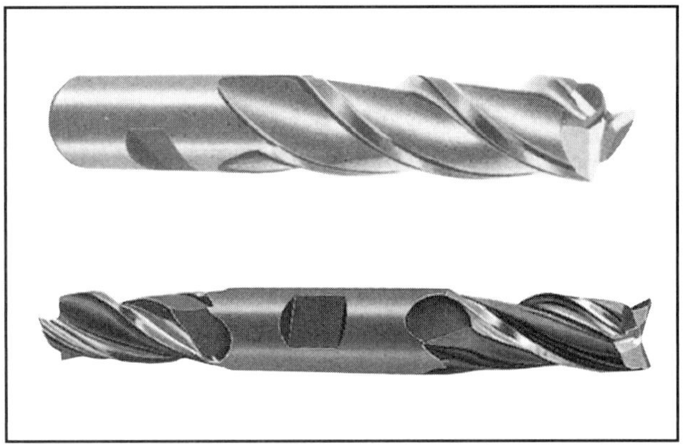

Figure 5.28 A single- (top) and double-ended (bottom) end mill. (Courtesy of Kennametal.)

Other considerations you will need to think about are the number of flutes, flute length, and end type. The type of material you will be cutting will dictate how many flutes you will need to machine the workpiece. The majority of end mills will have two or four flutes.

Number of Flutes

Single- or mono-flute end mills are used where high chip loads are to be expected. These end mills also require high rpms and are typically used on light alloy materials such as aluminum and plastic.

Two-flute end mills, like single-flute types, are used where fast chip removal is required. Three-flute end mills are typically found in larger diameters and will have a reduced shank. They can be used at moderate speeds on aluminum alloys and plastics, or at low-medium speeds for steel.

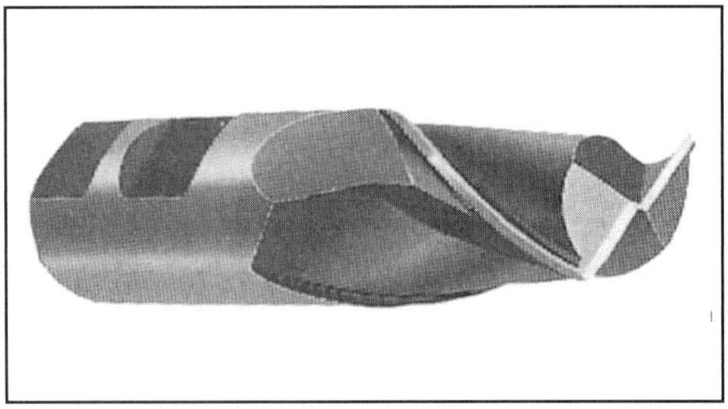

Figure 5.29 A two-flute end mill. (Courtesy of Kennametal.)

Two-Flute Single-End Mills

Mill Diameter	Shank Diameter	Flute Length	OAL
$^{1}/_{8}$"	$^{3}/_{8}$"	$^{3}/_{8}$"	$2^{5}/_{16}$"
$^{5}/_{32}$"	$^{3}/_{8}$"	$^{3}/_{8}$"	$2^{5}/_{16}$"
$^{3}/_{16}$"	$^{3}/_{8}$"	$^{3}/_{8}$"	$2^{5}/_{16}$"
$^{7}/_{32}$"	$^{3}/_{8}$"	$^{3}/_{8}$"	$2^{5}/_{16}$"
$^{1}/_{4}$"	$^{3}/_{8}$"	$^{1}/_{2}$"	$2^{5}/_{16}$"
$^{9}/_{32}$"	$^{3}/_{8}$"	$^{1}/_{2}$"	$2^{5}/_{16}$"
$^{5}/_{16}$"	$^{3}/_{8}$"	$^{9}/_{16}$"	$2^{5}/_{16}$"
$^{11}/_{32}$"	$^{3}/_{8}$"	$^{9}/_{16}$"	$2^{5}/_{16}$"
$^{3}/_{8}$"	$^{3}/_{8}$"	$^{9}/_{16}$"	$2^{5}/_{16}$"
$^{13}/_{32}$"	$^{3}/_{8}$"	$^{13}/_{16}$"	$2^{1}/_{2}$"
$^{7}/_{16}$"	$^{3}/_{8}$"	$^{13}/_{16}$"	$2^{1}/_{2}$"
$^{15}/_{32}$"	$^{3}/_{8}$"	$^{13}/_{16}$"	$2^{1}/_{2}$"
$^{1}/_{2}$"	$^{3}/_{8}$"	$^{13}/_{16}$"	$2^{1}/_{2}$"
$^{1}/_{2}$"	$^{1}/_{2}$"	1"	3"
$^{17}/_{32}$"	$^{1}/_{2}$"	$1^{1}/_{8}$"	3"
$^{9}/_{16}$"	$^{1}/_{2}$"	$1^{1}/_{8}$"	$3^{1}/_{8}$"
$^{5}/_{8}$"	$^{1}/_{2}$"	$1^{1}/_{8}$"	$3^{1}/_{8}$"
$^{5}/_{8}$"	$^{5}/_{8}$"	$1^{5}/_{16}$"	$3^{7}/_{16}$"
$^{11}/_{16}$"	$^{1}/_{2}$"	$1^{5}/_{16}$"	$3^{5}/_{16}$"
$^{11}/_{16}$"	$^{5}/_{8}$"	$1^{5}/_{16}$"	$3^{7}/_{16}$"
$^{3}/_{4}$"	$^{1}/_{2}$"	$1^{5}/_{16}$"	$3^{5}/_{16}$"
$^{3}/_{4}$"	$^{5}/_{8}$"	$1^{5}/_{16}$"	$3^{7}/_{16}$"
$^{3}/_{4}$"	$^{3}/_{4}$"	$1^{5}/_{16}$"	$3^{9}/_{16}$"
$^{13}/_{16}$"	$^{5}/_{8}$"	$1^{1}/_{2}$"	$3^{5}/_{8}$"
$^{13}/_{16}$"	$^{3}/_{4}$"	$1^{1}/_{2}$"	$3^{3}/_{4}$"
$^{7}/_{8}$"	$^{1}/_{2}$"	$1^{1}/_{2}$"	$1^{1}/_{2}$"
$^{7}/_{8}$"	$^{5}/_{8}$"	$1^{1}/_{2}$"	$3^{5}/_{8}$"
$^{7}/_{8}$"	$^{3}/_{4}$"	$1^{1}/_{2}$"	$3^{3}/_{4}$"
$^{7}/_{8}$"	$^{7}/_{8}$"	$1^{1}/_{2}$"	$3^{3}/_{4}$"

Figure 5.30 A sample of the commercially available two-flute single-end mills in sizes ranging from $^{1}/_{8}$" to $^{7}/_{8}$" diameter.

Two-Flute Double-End Mills

Mill Diameter	Shank Diameter	Flute Length	OAL
1/8"	3/8"	3/8"	3 1/16"
5/32"	3/8"	7/16"	3 1/8"
3/16"	3/8"	1/2"	3 1/4"
7/32"	3/8"	9/16"	3 1/4"
1/4"	3/8"	5/8"	3 3/8"
9/32"	3/8"	11/16"	3 3/8"
5/16"	3/8"	3/4"	3 1/2"
11/32"	3/8"	3/4"	3 1/2"
3/8"	3/8"	3/4"	3 1/2"
13/32"	1/2"	1"	4 3/16"
7/16"	1/2"	1"	4 1/8"
15/32"	1/2"	1"	4 1/8"
1/2"	1/2"	1"	4 1/8"
9/16"	5/8"	1 3/8"	5"
5/8"	5/8"	1 3/8"	5"
11/16"	3/4"	1 3/8"	5 5/8"
3/4"	3/4"	1 3/8"	5 5/8"
13/16"	7/8"	1 7/8"	6 1/8"
7/8"	7/8"	1 7/8"	6 1/8"

Figure 5.31 *A sample of the commercially available two-flute double-end mills in sizes ranging from 1/8" to 7/8" diameter.*

Four-flute end mills are regularly used for steel. End mills with more than four flutes will generally have a reduced shank and are used for all material types.

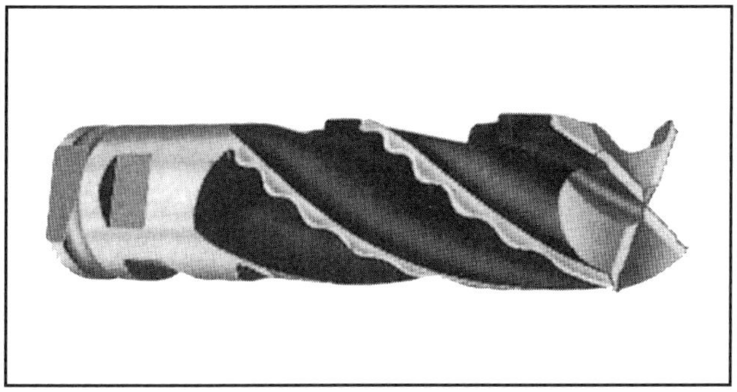

Figure 5.32 *A four-flute end mill. (Courtesy of Kennametal.)*

To improve the rate at which material is removed through the flutes, some end mills have a low or high *helix* angle. The helix is the spiral angle of the flute. A low-helix is the most common type found on end mills. High-helix end mills are used for high speeds on aluminum and high temperature materials. A *reverse* helix end mill is used for cutting thin sheet stock.

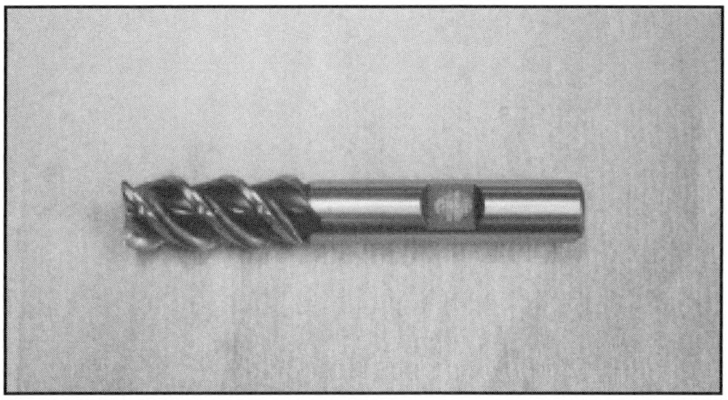

Figure 5.33 *Helix refers to the amount of "twist" placed on the flutes.*

The manner in which the end mill engages the workpiece is somewhat determined by the configuration on the bottom of the cutting edge. The cutting edge will have either a small diameter hole in it, or it will be *center cutting*. Center cutting end mills can be used much like a drill bit because they can drill (a loose term here) a hole when they are fed straight down. This can be an important plus if you are *pocket* or *frame* milling. The end mill can be positioned and fed straight down into the workpiece, then moved from position to position. Non-center cutting end mills will feed down until the center hole is filled up with material, at which time the downward movement will stop.

Figure 5.34 *Center cutting end mills can be used to bore straight into the workpiece.*

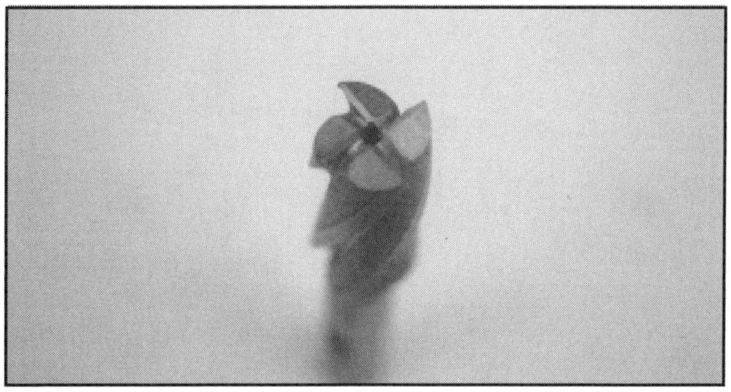

Figure 5.35 *This four-flute end mill has a hole in the middle of the flutes which prevents it from machining a hole straight down into a workpiece.*

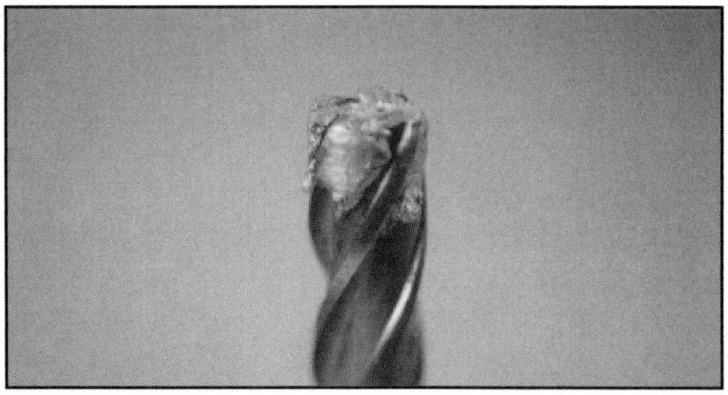

Figure 5.36 *This four-flute end mill is a non-center cutting type and was used bore a hole. As you can see, the hole in the center prevented the completion of the hole. The hole was also oversize and had a good amount of taper. Both part and end mill were scrapped.*

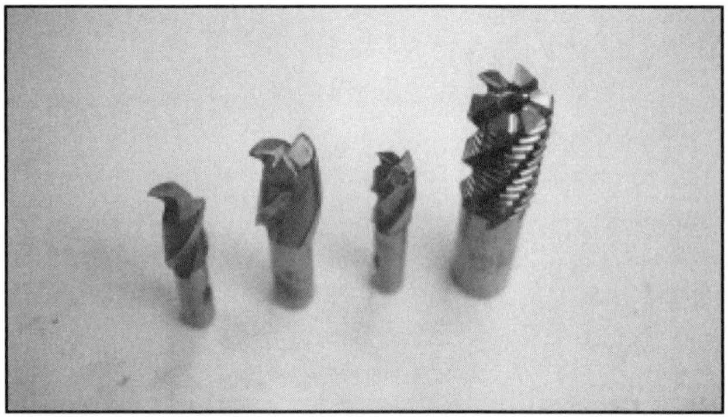

Figure 5.37 *2-flute, 3-flute, 4-flute, and 6-flute end mills.*

Other End Mill Types

There is a large variety of end mills available for specialized applications and machining routines. They allow different shapes to be cut into the workpiece. There are times when slots, angles, rounded corners, keyways, and dovetails must be machined. We will examine these various types and look at some of their physical attributes.

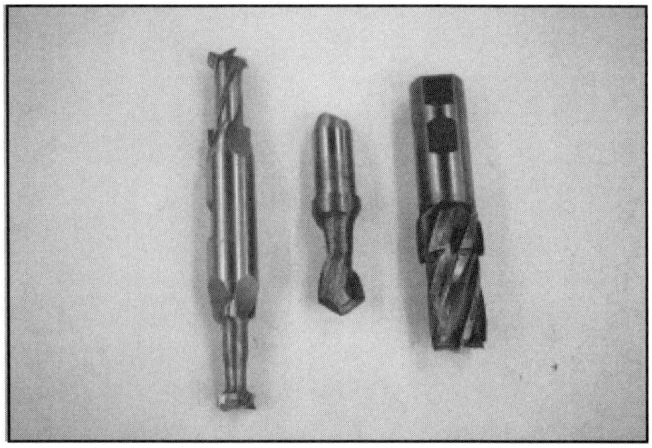

Figure 5.38 *These end mills have been ground with special sizes and shapes to cut a profile into the surface of a workpiece.*

Ball End Mills

Ball end mills are used to cut a half-round (hemisphere) shape into the surface of a workpiece. This type of end mill is also used where a radius is required to connect two flat and/or angled flat surfaces. A radius connection between two flat surfaces is machined to reduce the possibility of *stress risers*, or for decorative purposes. The majority of ball end mills are made from high-speed steel, with or without TiN coatings. When specifying sizes, do not indicate a radius, only the diameter. Ball end mills will have two or four flutes and have an RH spiral for RH cutting.

Ball End Mills

Cutter Diameter	Shank Diameter	Length of Cut	OAL
1/16"	1/8"	3/16"	1 1/2"
3/32"	1/8"	3/8"	1 1/2"
1/8"	1/8"	1/2"	1 1/2"
5/32"	3/16"	9/16"	2"
3/16"	3/16"	5/8"	2"
7/32"	1/4"	5/8"	2 1/2"
1/4"	1/4"	3/4"	2 1/2"
9/32"	5/16"	3/4"	2 1/2"
5/16"	5/16"	13/16"	2 1/2"
11/32"	3/8"	7/8"	2 1/2"
3/8"	3/8"	7/8"	2 1/2"
13/32"	7/16"	7/8"	2 1/2"
7/16"	7/16"	7/8"	2 1/2"
15/32"	1/2"	1"	3"
1/2"	1/2"	1"	3"
9/16"	9/16"	1 1/4"	3"
5/8"	5/8"	1 1/4"	3 1/2"
3/4"	3/4"	1 1/2"	4"
7/8"	7/8"	1 1/2"	4"
1"	1"	1 1/2"	4"

Figure 5.39 *Typical specifications of ball end mills commercially available.*

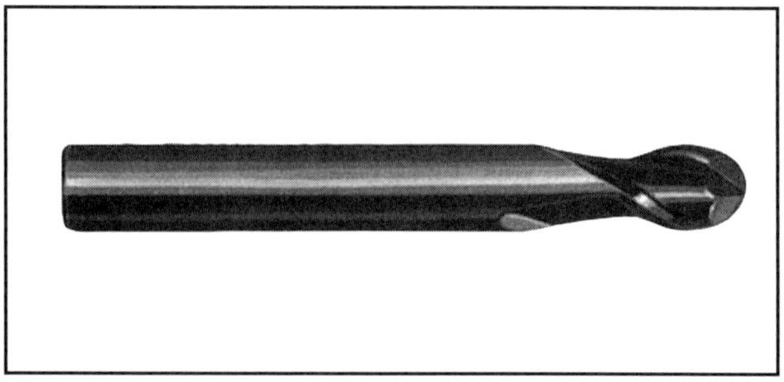

Figure 5.40 *A two-flute ball end mill. (Courtesy of Kennametal.)*

Roughing End Mills

Roughing end mills have a very different flute design than straight fluted types. They have a continuous spiral cut into the edge of the flute to allow coolant into the cutting area. This design also improves chip removal from the workpiece. Roughing end mills will have no fewer than three flutes and generally no more than eight. They are used where high production rates are the norm. Cutting of space-age metals like Inconel, Hastalloy, titanium, and certain stainless steel alloys also benefit from this cutter design. Increased tool life and reduced vibration are some of the most important reasons to use roughing end mills. Sizes range from $^{1}/_{4}$" to 2" diameter, and lengths from $2^{7}/_{16}$" to $11^{3}/_{4}$". They are available with or without TiN coatings.

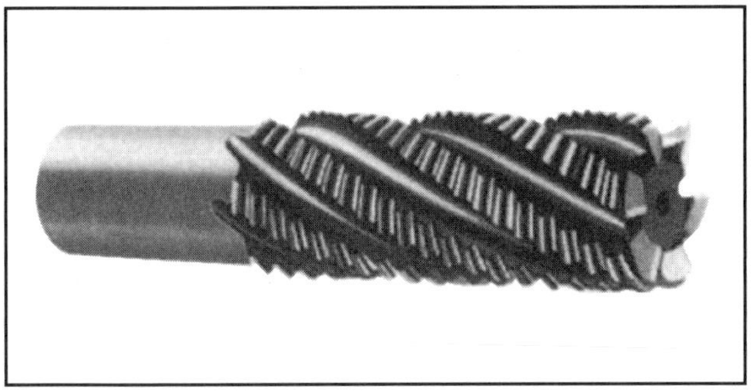

Figure 5.41 *A roughing end mill.*

Roughing End Mills

Cutter Diameter	Shank Diameter	Length of Cut	OAL
$^{1}/_{4}$"	$^{3}/_{8}$"	$^{5}/_{8}$"	$2^{7}/_{16}$"
$^{5}/_{16}$"	$^{3}/_{8}$"	$^{3}/_{4}$"	$2^{9}/_{16}$"
$^{3}/_{8}$"	$^{3}/_{8}$"	$^{3}/_{4}$"	$2^{9}/_{16}$"
$^{7}/_{16}$"	$^{3}/_{8}$"	1"	$2^{11}/_{16}$"
$^{1}/_{2}$"	$^{1}/_{2}$"	$1^{1}/_{4}$"	$3^{1}/_{4}$"
$^{1}/_{2}$"	$^{1}/_{2}$"	2"	4"
$^{5}/_{8}$"	$^{5}/_{8}$"	$1^{5}/_{8}$"	$3^{3}/_{4}$"
$^{5}/_{8}$"	$^{5}/_{8}$"	$2^{1}/_{2}$"	$4^{5}/_{8}$"
$^{3}/_{4}$"	$^{3}/_{4}$"	$^{3}/_{4}$"	3"
$^{3}/_{4}$"	$^{3}/_{4}$"	1-$^{5}/_{8}$"	$3^{7}/_{8}$"
$^{3}/_{4}$"	$^{3}/_{4}$"	3"	$5^{1}/_{4}$"
$^{7}/_{8}$"	$^{3}/_{4}$"	$3^{1}/_{2}$"	$5^{3}/_{4}$"
1"	$^{3}/_{4}$"	$^{3}/_{4}$"	3"
1"	$^{3}/_{4}$"	$1^{1}/_{2}$"	$3^{3}/_{4}$"
1"	1"	2"	$4^{1}/_{2}$"
1"	1"	3"	$5^{1}/_{2}$"
1"	1"	4"	$6^{1}/_{2}$"
1"	1"	6"	$8^{1}/_{2}$"

Figure 5.42 *Typical specifications of roughing end mills commercially available.*

Tapered End Mills

Tapered end mills are used primarily by pattern and die makers who must cut *draft angles* onto the edge or sides of their workpiece. This allows the pattern or part to be removed from the mold or die with less effort.

Tapered end mills are manufactured in a wide variety of angles and flute lengths to handle the various materials used in mold and die making. Most typical of the flute types are the *right hand* (RH) cut with a *right hand* spiral. Tapers, which are commercially available, include: $^{1}/_{2}$°, 1°, $1^{1}/_{2}$°, 2°, 3°, 5°, 7°, 10°, 15°, 20°, and 25° per side. Specified angles are *per side* — not the *included angle,* e.g., a $^{1}/_{2}$° angled cutter would cut a $^{1}/_{2}$° angle onto the edge of the workpiece, or a 1° included angle down the middle.

Figure 5.43 *This workpiece is being machined with a 3° four-flute tapered end mill.*

Tip diameters vary in size from .070" to $^{3}/_{4}$" depending upon the tool's length. Tool lengths will vary from $2^{1}/_{2}$" to $7^{3}/_{4}$" depending upon the angle of the cutter.

Tapered End Mills with 1/2° Per Side

Taper (per side)	Tip Diameter	Flute Length	Shank Diameter	OAL
1/2°	3/32"	1/2"	3/8"	2 1/2"
	3/32"	3/4"	3/8"	2 1/2"
	3/32"	1 1/4"	3/8"	3 1/8"
	1/8"	1/4"	3/8"	2 1/2"
	1/8"	1/2"	3/8"	2 1/2"
	1/8"	3/4"	3/8"	2 1/2"
	1/8"	1"	3/8"	2 7/8"
	1/8"	1 1/4"	3/8"	3 1/8"
	3/16"	3/4"	3/8"	2 1/2"
	3/16"	1 1/4"	3/8"	3 1/8"
	1/4"	3/4"	3/8"	2 1/2"
	1/4"	1 1/4"	3/8"	3 1/8"
	1/4"	2 1/4"	1/2"	4 1/4"
	1/4"	3 1/4"	1/2"	5 1/2"
	3/8"	2 1/4"	1/2"	4 1/4"
	3/8"	3 1/4"	1/2"	5 1/2"
	1/2"	1 1/4"	1/2"	3 1/8"
	1/2"	2 1/4"	1/2"	4 1/4"
	1/2"	3 1/4"	1/2"	5 1/2"
	5/8"	2 1/4"	5/8"	4 1/2"
	5/8"	3 1/4"	5/8"	5 1/2"
	5/8"	4 1/4"	5/8"	6 1/2"
	3/4"	2 1/4"	3/4"	4 1/2"
	3/4"	3 1/4"	3/4"	5 3/4"
	3/4"	4 1/4"	3/4"	7 3/4"

Figure 5.44 *Chart for commercially available 1/2° per side tapered end mills.*

Tapered End Mills with 1° Per Side

Taper (per side)	Tip Diameter	Flute Length	Shank Diameter	OAL
1°	3/32"	1/2"	3/8"	2 1/2"
	3/32"	3/4"	3/8"	2 1/2"
	3/32"	1-1/4"	3/8"	3 1/8"
	1/8"	1/4"	3/8"	2 1/2"
	1/8"	1/2"	3/8"	2 1/2"
	1/8"	3/4"	3/8"	2 1/2"
	1/8"	1"	3/8"	2 7/8"
	1/8"	1 1/4"	3/8"	3 1/8"
	3/16"	3/4"	3/8"	2 1/2"
	3/16"	1 1/4"	3/8"	3 1/8"
	1/4"	3/4"	3/8"	2 1/2"
	1/4"	1 1/4"	3/8"	3 1/8"
	1/4"	2 1/4"	1/2"	4 1/4"
	1/4"	3 1/4"	1/2"	5 1/2"
	3/8"	2 1/4"	1/2"	4 1/4"
	3/8"	3 1/4"	1/2"	5 1/2"
	1/2"	1 1/4"	1/2"	3 1/8"
	1/2"	2 1/4"	1/2"	4 1/4"
	1/2"	3 1/4"	1/2"	5 1/2"

Tapered End Mills with 1° Per Side *(continued)*

Taper (per side)	Tip Diameter	Flute Length	Shank Diameter	OAL
1°	5/8"	2 1/4"	5/8"	4 1/2"
	5/8"	3 1/4"	5/8"	5 1/2"
	5/8"	4 1/4"	5/8"	6 1/2"
	3/4"	2 1/4"	3/4"	4 1/2"
	3/4"	3 1/4"	3/4"	5 3/4"
	3/4"	4 1/4"	3/4"	6 3/4"

Figure 5.45 *Chart for commercially available 1° per side tapered end mills.*

Tapered End Mills with 1 1/2° Per Side

Taper (per side)	Tip Diameter	Flute Length	Shank Diameter	OAL
1-1/2°	3/32"	1/2"	3/8"	2 1/2"
	3/32"	3/4"	3/8"	2 1/2"
	3/32"	1 1/4"	3/8"	3 1/8"
	7/32"	1"	3/8"	2 7/8"
	1/8"	1/4"	3/8"	2 1/2"
	1/8"	1/2"	3/8"	2 1/2"
	1/8"	3/4"	3/8"	2 1/2"
	1/8"	1"	3/8"	2 7/8"
	1/8"	1 1/4"	3/8"	3 1/8"
	3/16"	3/4"	3/8"	2 1/2"
	3/16"	1 1/4"	3/8"	3 1/8"
	1/4"	3/4"	3/8"	2 1/2"
	1/4"	1 1/4"	3/8"	3 1/8"
	1/4"	2 1/4"	1/2"	4 1/4"
	1/4"	3 1/4"	1/2"	5 1/2"
	3/8"	2 1/4"	1/2"	4 1/4"
	3/8"	3 1/4"	1/2"	5 1/2"
	1/2"	1 1/4"	1/2"	3 1/8"
	1/2"	2 1/4"	5/8"	4 1/4"
	1/2"	3 1/4"	5/8"	5 1/2"

Figure 5.46 *Chart for commercially available 1 1/2° per side tapered end mills.*

Tapered End Mills with 2° Per Side

Taper (per side)	Tip Diameter	Flute Length	Shank Diameter	OAL
2°	3/32"	1/2"	3/8"	2 1/2"
	3/32"	3/4"	3/8"	2 1/2"
	1/8"	1/2"	3/8"	2 1/2"
	1/8"	3/4"	3/8"	2 1/2"
	1/8"	1"	3/8"	2 1/2"
	1/8"	1 1/4"	3/8"	3 1/8"
	3/16"	3/4"	3/8"	2 1/2"
	3/16"	1 1/4"	3/8"	3 1/8"
	1/4"	3/4"	3/8"	2 1/2"
	1/4"	1 1/4"	3/8"	3 1/8"
	1/4"	2 1/4"	1/2"	4 1/4"
	1/4"	3 1/4"	1/2"	5 1/2"
	3/8"	1 1/4"	1/2"	2 1/2"
	3/8"	2 1/4"	1/2"	4 1/4"
	3/8"	3 1/4"	3/4"	5 1/2"
	1/2"	1 1/4"	1/2"	3 1/8"
	1/2"	2 1/4"	1/2"	4 1/4"
	1/2"	3 1/4"	3/4"	5 1/2"

Figure 5.47 *Chart for commercially available 2° per side tapered end mills.*

Tapered End Mills with 3° Per Side

Taper (per side)	Tip Diameter	Flute Length	Shank Diameter	OAL
3°	3/32"	1"	3/8"	2 7/8"
	3/32"	1 1/4"	3/8"	3 1/8"
	3/32"	1 1/2"	3/8"	3 1/4"
	3/32"	2"	3/8"	3 5/8"
	3/32"	2 1/2"	1/2"	4 1/2"
	7/64"	1"	3/8"	2 7/8"
	7/64"	2"	3/8"	3 5/8"
	1/8"	3/4"	3/8"	2 1/2"
	1/8"	1"	3/8"	2 7/8"
	1/8"	1 1/4"	3/8"	3 1/8"
	1/8"	1 1/2"	3/8"	3 1/4"
	3/16"	3/4"	3/8"	2 1/2"
	3/16"	1 1/4"	3/8"	3 1/8"
	3/16"	3 1/4"	3/8"	3 1/4"
	1/4"	3/4"	3/8"	2 1/2"
	1/4"	1 1/4"	3/8"	3 1/8"
	1/4"	2 1/4"	1/2"	4 1/4"
	3/8"	1 1/4"	1/2"	3 1/8"
	3/8"	2 1/4"	1/2"	4 1/4"
	3/8"	3 1/4"	3/4"	5 1/2"
	1/2"	1 1/4"	1/2"	3 1/8"
	1/2"	2 1/4"	3/4"	4 1/4"

Figure 5.48 *Chart for commercially available 3° per side tapered end mills.*

Tapered End Mills with 5° Per Side

Taper (per side)	Tip Diameter	Flute Length	Shank Diameter	OAL
5°	3/32"	3/4"	3/8"	2 1/2"
	3/32"	1"	3/8"	2 1/2"
	3/32"	1 1/4"	3/8"	
	3/32"	1 1/2"	3/8"	
	3/32"	2"	1/2"	
	7/64"	1 1/2"	3/8"	
	7/64"	2"	1/2"	
	1/8"	3/4"	3/8"	2 1/2"
	1/8"	1"	3/8"	2 1/2"
	1/8"	1 1/4"	3/8"	2 1/2"
	1/8"	2"	1/2"	3 1/8"
	3/16"	3/4"	3/8"	2 1/2"
	3/16"	1 1/4"	3/8"	3 1/8"
	3/16"	2 1/2"	1/2"	
	3/16"	3 1/4"	3/4"	
	1/4"	3/4"	3/8"	2 1/2"
	1/4"	1 1/4"	1/2"	3 1/8"
	1/4"	2 1/4"	1/2"	4 1/4"
	1/4"	3 1/4"	3/4"	5 1/2"
	3/8"	1 1/4"	1/2"	2 1/2"
	3/8"	2 1/4"	5/8"	4 1/4"
	3/8"	3 1/4"	3/4"	5 1/2"
	1/2"	1 1/4"	1/2"	3 1/8"
	1/2"	2 1/4"	3/4"	4 1/4"

Figure 5.49 *Chart for commercially available 5° per side tapered end mills.*

Tapered End Mills with 7° Per Side

Taper (per side)	Tip Diameter	Flute Length	Shank Diameter	OAL
7°	.070"	1 1/2"	1/2"	2 1/2"
	3/32"	1"	3/8"	2 1/2"
	3/32"	1 1/4"	1/2"	2 1/2"
	3/32"	1 1/2"	1/2"	
	1/8"	3/4"	3/8"	2 1/2"
	1/8"	1"	3/8"	2 1/2"
	1/8"	1 1/2"	1/2"	3 1/8"
	3/16"	1 1/4"	1/2"	3 1/8"
	1/4"	3/4"	1/2"	3 1/8"
	1/4"	1 1/4"	1/2"	4 1/4"
	1/4"	2 1/4"	3/4"	5 1/2"
	3/8"	2 1/4"	3/4"	2 1/2"
	1/2"	1 1/4"	5/8"	4 1/4"

Figure 5.50 *Chart for commercially available 7° per side tapered end mills.*

Tapered End Mills with 10° Per Side

Taper (per side)	Tip Diameter	Flute Length	Shank Diameter	OAL
10°	3/32"	1/2"	3/8"	2 1/2"
	3/32"	1 1/2"	1/2"	2 1/2"
	1/8"	3/4"	3/8"	2 1/2"
	1/8"	1 1/4"	1/2"	2 1/2"
	3/16"	1 1/4"	1/2"	3 1/8"
	1/4"	3/4"	1/2"	3 1/8"
	1/4"	1 1/4"	1/2"	4 1/4"
	1/4"	2 1/4"	3/4"	5 1/2"
	3/8"	1 1/4"	5/8"	2 1/2"
	3/8"	2 1/4"	1"	4 5/8"
	1/2"	2 1/4"	3/4"	3 3/8"
	1/2"	2 1/4"	1"	5 5/8"

Figure 5.51 *Chart for commercially available 10° per side tapered end mills.*

Tapered End Mills with 15° Per Side

Taper (per side)	Tip Diameter	Flute Length	Shank Diameter	OAL
15°	3/32"	3/4"	1/2"	2 1/2"
	3/32"	1"	1/2"	2 1/2"
	3/32"	1 1/2"	1/2"	
	1/8"	1/2"	3/8"	2 1/2"
	1/8"	3/4"	1/2"	2 1/2"
	1/8"	1"	1/2"	2 1/2"
	1/8"	1 1/4"	1/2"	3 1/8"
	3/16"	1"	1/2"	2 1/2"
	3/16"	1 1/4"	1/2"	3 1/8"
	3/16"	1 1/4"	3/4"	
	1/4"	1"	1/2"	2 1/2"
	1/4"	1 1/4"	3/4"	3 1/8"
	1/4"	1 1/4"	1/2"	4 1/4"
	1/4"	1 1/2"	3/4"	5 1/2"
	3/8"	1 1/4"	3/4"	2 1/2"
	1/2"	1"	3/4"	3 1/8"
	1/2"	1 1/4"	3/4"	4 1/4"

Figure 5.52 *Chart for commercially available 15° per side tapered end mills.*

Tapered End Mills with 20° Per Side

Taper (per side)	Tip Diameter	Flute Length	Shank Diameter	OAL
20°	1/8"	1/2"	3/8"	2 1/2"
	1/8"	1"	1/2"	3"
	3/16"	1"	1/2"	3"
	1/4"	1"	1/2"	3"
	5/16"	1"	3/4"	3 1/4"

Figure 5.53 *Chart for commercially available 20° per side tapered end mills.*

Tapered End Mills with 25° Per Side

Taper (per side)	Tip Diameter	Flute Length	Shank Diameter	OAL
25°	1/8"	1/2"	1/2"	2 1/2"
	1/8"	1"	1/2"	3"
	1/4"	1"	5/8"	3 1/4"
	5/16"	3/4"	5/8"	3"
	5/16"	1"	3/4"	3 1/4"
	3/8"	3/4"	5/8"	3"

Figure 5.54 *Chart for commercially available 25° per side tapered end mills.*

Dovetail Cutters

Dovetail cutters are used to cut a dovetail angle onto/into a workpiece. Dovetails are used to align sliding parts so that they track along the same plane or event line in opposite directions. The majority of all machines use a dovetail when two components must slide back and forth atop one another, such as the mill table atop the knee. Cutters come in two angles — 45° and 60°. The particular angle you choose must be cut into both workpieces so that they will match one another.

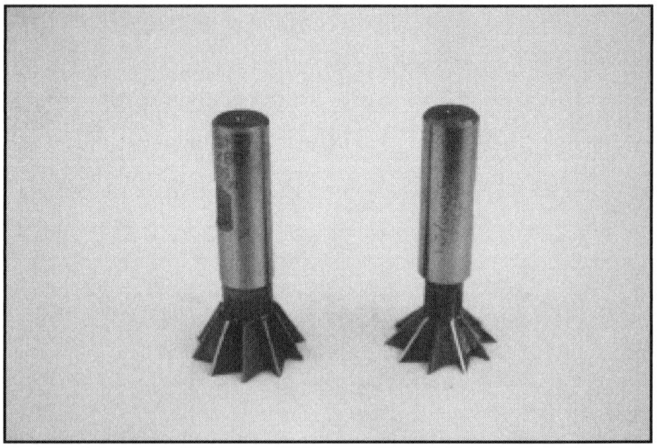

Figure 5.55 *A pair of dovetail cutters (left, 60°; and right, 45°).*

Dovetail Cutters

	Cutter Width			
Dia.	45°	60°	Shank Dia.	OAL
3/8"	1/8"	3/16"	3/8"	2 1/8"
1/2"	5/32"	1/4"	3/8"	2 1/8"
3/4"	1/4"	5/16"	3/8"	2 1/8"
1 3/8"	7/16"	5/8"	5/8"	2 7/8"
1 7/8"	5/8"	13/16"	7/8"	3 1/4"
2 1/4"	1 1/16"	1 1/16"	1"	3 3/4"

Figure 5.56 *Specifications of typical dovetail cutter commercially available.*

Dovetails are <u>not</u> typically cut in one pass with the cutter. Start by cutting a slot into the workpiece with a straight-fluted end mill and then finishing with the dovetail cutter. Great care must be taken **not** to cut the widths of the internal and external dovetails too wide or narrow.

T-Slot Cutters

These cutters are primarily used for cutting *T-slots* in tables and beds of machine tools. Like dovetail cutters, they require a vertical slot to be cut into the workpiece before the slot can be machined. This is a useful tool when you must make your own tooling and fixtures. T-slot nuts can be purchased in a wide variety of standardized sizes to fit the newly cut T-slots.

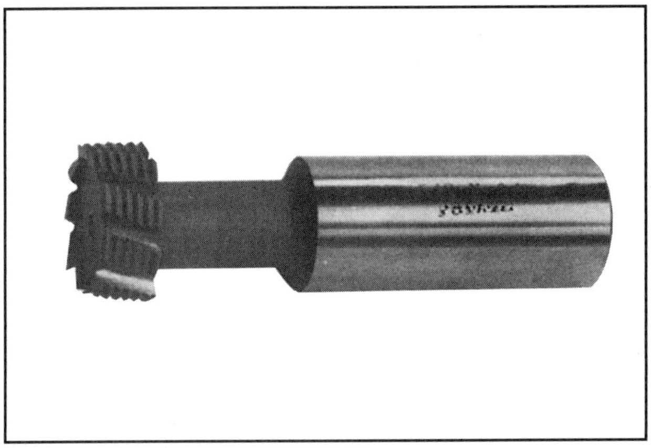

Figure 5.57 *A T-slot cutter.*

Cutters are supplied in roughing and finish types. Careful attention to coolant selection and usage is important in T-slot cutting. Remove excessive chip build-up with brushes or compressed air during the machining procedure. Chip build-up can cause the T-slot to become oversized as the cutter recuts the same chips.

T-Slot Cutters

Bolt Size	Cutter Diameter	Shank Diameter	Cutter Width	OAL
1/4"	9/16"	1/2"	15/64"	2 19/32"
5/16"	21/32"	1/2"	17/64"	2 11/16"
3/8"	25/32"	3/4"	21/64"	3 1/4"
1/2"	31/32"	3/4"	25/64"	3 7/16"
5/8"	1 1/4"	1"	31/64"	3 15/16"
3/4"	1 15/32"	1"	5/8"	4 7/16"
1"	1 27/32"	1 1/4"	53/64"	4 13/16"
1 1/4"	2 7/32"	1 1/4"	1 3/32"	5 3/8"
1 1/2"	2 21/32"	1 1/4"	1 11/32"	5 29/32"

Figure 5.58 *Specifications of typical T-slot cutters commercially available.*

Corner Rounding End Mills

When an external rounded edge is to be machined onto the workpiece, use a *corner rounding* end mill. These end mills have a specified <u>radius</u> cut into them and typically have a reduced diameter shank.

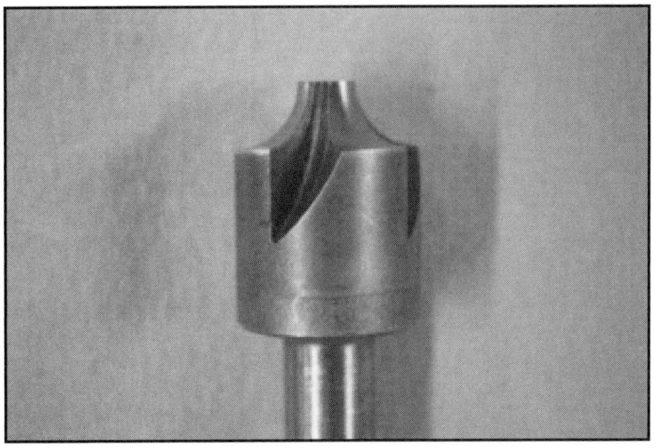

Figure 5.59 *A corner rounding end mill.*

Corner Rounding End Mills

Radius	Shank Diameter	Largest Diameter	OAL
1/16"	3/8"	7/16"	2 1/2"
3/32"	3/8"	1/2"	2 1/2"
1/8"	1/2"	5/8"	3"
5/32"	1/2"	3/4"	4"
3/16"	1/2"	7/8"	3 1/8"
3/16"	3/4"	7/8"	3 1/8"
1/4"	1/2"	1"	3 1/4"
1/4"	3/4"	1"	3 1/4"
5/16"	1/2"	1 1/8"	3 1/4"
5/16"	7/8"	1 1/8"	3 1/2"
3/8"	1/2"	1 1/4"	3 1/2"
3/8"	7/8"	1 1/4"	3 3/4"
7/16"	3/4"	1 3/8"	3 3/4"
7/16"	1"	1 3/8"	4"
1/2"	3/4"	1 1/2"	3 7/8"
1/2"	1"	1 1/2"	4 1/8"
3/4"	3/4"	2"	4"

Figure 5.60 *Specifications of typical corner rounding cutters commercially available.*

Radius Shank Cutters

Radius shank cutters are used for edge or interior machining of the workpiece. They are manufactured with concave and convex shapes. The concave shape can cut a radius onto the outer edge of the workpiece without having to turn the piece over as would be the case with a corner-rounding cutter.

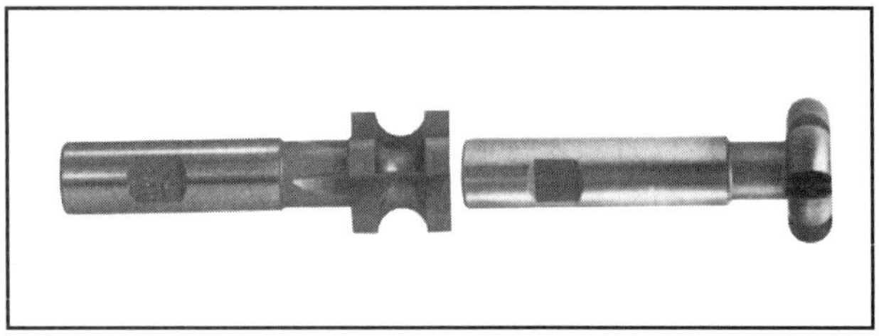

Figure 5.61 *A pair of radius shank cutters. Concave (left), convex (right).*

Concave Radius Shank Cutters

Radius	Shank Diameter	Cutter Diameter	OAL
1/32"	1/2"	3/4"	3"
1/16"	1/2"	3/4"	3"
3/32"	1/2"	3/4"	3"
1/8"	3/4"	1"	3 1/2"
5/32"	3/4"	1"	3 1/2"
3/16"	3/4"	1"	3 1/2"
1/4"	3/4"	1 1/4"	4"
5/16"	3/4"	1 1/4"	4"
3/8"	3/4"	1 1/2"	4 3/16"

Figure 5.62 *Specification of concave radius shank cutters commercially available.*

Convex cutters are used to machine a half circle into the surface edge of a workpiece. Such a shape may be used for aesthetic or functional design purposes.

Convex Radius Shank Cutters

Radius	Shank Diameter	Cutter Diameter	OAL
1/32"	1/2"	3/4"	3"
1/16"	1/2"	3/4"	3"
3/32"	1/2"	7/8"	3"
1/8"	3/4"	1 1/4"	3 1/2"
5/32"	3/4"	1 5/16"	3 1/2"
3/16"	3/4"	1 3/8"	3 1/2"
1/4"	3/4"	1 1/2"	4"
5/16"	3/4"	1 5/8"	4"
3/8"	3/4"	1 3/4"	4"

Figure 5.63 *Specifications of convex radius shank cutters commercially available.*

Double Angle Shank Cutters

The *double angle shank cutter* is used for either chamfering edges, or cutting a V-groove detail into a workpiece in a single pass. They are manufactured from high-speed steel and have a 60° or 90° included angle. Their use on a vertical milling machine will require the use of a stub arbor, or a right angle milling attachment.

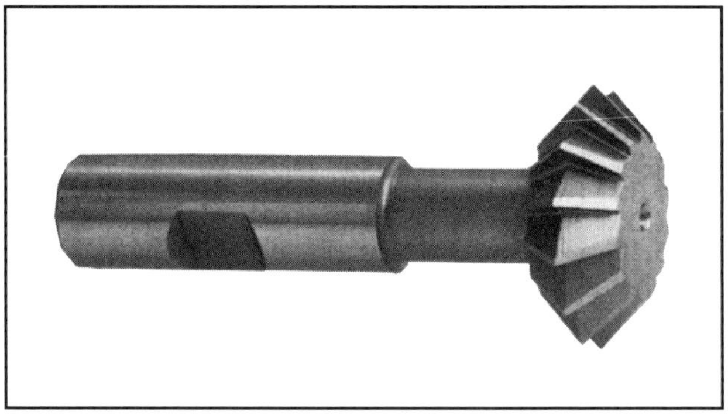

Figure 5.64 *A double angle shank cutter.*

Double Angle Shank Cutters w/60° (Included)

Shank Diameter	Cutter Diameter	OAL	Cutter Width
3/8"	3/4"	2 3/8"	3/16"
1/2"	1"	2 27/32"	5/16"
5/8"	1 3/8"	3 7/32"	7/16"
5/8"	1 1/2"	3 3/8"	1/2"
3/4"	1 7/8"	3 25/32"	5/8"
3/4"	2 1/4"	3 1/8"	1/2"
7/8"	2 1/4"	4 5/32"	3/4"

Figure 5.65 *Specifications of 60° (included angle) double angle shank cutters commercially available.*

Double Angle Shank Cutters w/90° (Included)

Shank Diameter	Cutter Diameter	OAL	Cutter Width
3/8"	3/4"	2 3/8"	1/4"
1/2"	1"	2 27/32"	3/8"
5/8"	1 3/8"	3 7/32"	1/2"
5/8"	1 1/2"	3 3/8"	9/16"
3/4"	1 7/8"	3 25/32"	5/8"
3/4"	2 1/4"	3 1/8"	1/2"
7/8"	2 1/4"	4 5/32"	3/4"

Figure 5.66 *Specifications of 90° (included angle) double angle shank cutters commercially available.*

Single Angle Chamfering Cutters

The *single angle chamfering* cutter is designed for one purpose — chamfering of the workpiece. These cutters find their greatest applications in production shops where tools are setup in quick-change tool changers. They are made from high-speed steel and are manufactured with a 45° or 60° angle.

Figure 5.67 *A single angle chamfering cutter.*

Single Angle Chamfering Cutters

Cutter Diameter	Cutter Width	Shank Diameter
1/2"	1/8"	3/8"
3/4"	3/16"	3/8"
1"	5/16"	1/2"
1-1/2"	1/2"	3/4"

Figure 5.68 *Specifications of single angle chamfering cutters commercially available.*

Plain Mill Cutters

Plain mill cutters are used to remove large amounts of material. Because of their size and typical horizontal mill operation, they find limited use on vertical milling machines. However, you need to understand their function and usability because they can be mounted onto *right-angle* adapters that will align them for horizontal operation.

Figure 5.69 *A plain mill cutter.*

Cutters are made from high-speed steel and have straight or spiraled teeth. These cutters are available in a wide variety of diameters and widths to suit various applications. Special sizes can be ordered to assist special machining jobs.

Plain mills can be *stacked* or *spaced* on the arbor to machine various shapes onto a workpiece. Cutters are typically fed straight into the work path because they may have no, or limited, side milling capabilities.

Figure 5.70 *This set of plain mill cutters (gang milling) is used to machine a shape onto the surface of a workpiece. This set up is on a horizontal mill.*

Shell Mills

Shell end mills are used to machine wide or flat areas onto the workpiece. They mount onto special shell mill arbors. A selection of arbors may be necessary to mount the cutters because of the different mounting-hole sizes available. Cutters are made from high-speed steel or made with steel bodies with cemented carbide tips. Production shops will generally use cutter bodies fitted with replaceable carbide inserts. Insert shapes include: triangle, square, diamond, and round types. You may also find shell mill bodies that accept square high-speed steel or carbide-tipped tool bits.

Figure 5.71 *A typical shell end mill; this one uses replaceable carbide-tipped tools.*

Solid-type shell mills will have four, six, eight, or more cutting edges depending upon workpiece material type and diameter. Cutters with fewer teeth (four or six) are used for light alloys to allow increased chip loads and provide additional chip removal clearance space. Roughing end mills are used for production and roughing applications. Tools can be specified with RH or LH tooth profiles.

Insert-type shell mills come in a wide variety of sizes and configurations. Often the cutter will not have a 90° insert profile, in order to eliminate *chatter* marks. Chatter is a resonant vibration caused when there is excessive contact between the cutter blade/tooth and the workpiece. Since chatter produces a poor surface finish, insert selection is important in order to achieve the correct surface finish.

> **!**
> *Do not overload your mill with too large a diameter shell mill. Damage to the spindle bearings may cause premature failure and problems with maintaining accuracy.*

Shell End Mills - Solid Type

Cutter Diameter	Flute Length	Mounting Hole Size
1 1/4"	1"	1/2"
1 1/2"	1 1/8"	1/2"
1 3/4"	1 1/4"	3/4"
2"	1 3/8"	3/4"
2 1/4"	1 1/2"	1"
2 1/2"	1 5/8"	1"
2 3/4"	1 5/8"	1"
3"	1 3/4"	1 1/4"
3 1/2"	1 7/8"	1 1/4"
4"	2 1/4"	1 1/2"
4 1/2"	2 1/4"	1 1/2"
5"	2 1/4"	1 1/2"
5 1/2"	2 1/4"	2"
6"	2 1/4"	2"

Figure 5.72 *Specifications of shell end mills commercially available.*

Metal Slitting Saws

This group of cutters is used for deep sawing, plunging, and slotting applications where narrow dimensions are required. They are designated as either *plain* or *side cutting*. For practicality reasons, saw diameters should be limited to 5" or less on vertical mills when the cutter is used in a horizontal (typical) cutting location. Cutters must be installed onto arbors for use. It is important to install the correct size keyway into the arbor before the cutter is installed, otherwise the cutter will slip during operation.

Figure 5.73 *A pair of metal slitting saws. Note the saw on top (left) is a side cutting type, the saw on the bottom (right) is a plain type.*

Plain metal slitting saws are used for long continuous slitting operations because they don't have *side clearance* for chips.

Side cutting slitting saws are used when plunge (deep) cutting into a part, or sawing operations.

Slitting saws are made from high-speed steel, or steel bodies with cemented carbide blades. High production shops will use insert-type slitting cutters.

Jeweler's Slotting Saws

Like their metal slitting cutter cousins, jeweler's slotting saws are used to cut narrow slots into sheeting or tubing workpieces. These saw blades come in very thin sizes, some as small as .008" for 1" diameter sizes and up to .128" thick for 6" diameters. Jeweler's saws are made from high-speed steel.

Figure 5.74 *A jeweler's slotting saw is similar in style to the slitting saw. Note that an angle plate is used to align the right angle adapter.*

Some jeweler's, saws are *hollow ground* to provide necessary side clearance during operation, otherwise they are machined with flat parallel sides.

Single Angle Cutters

When a workpiece requires a chamfer of 45° or 60° angle, a single angle cutter works very well. These cutters are made of high-speed steel and come in a variety of diameters. The cutters are typically mounted onto an arbor. Mounted in a vertical mill, the cutter is stationed in a horizontal position.

Figure 5.75 *Single angle cutters are available in 45° and 60° cutting surfaces.*

Single Angle Cutters

Diameter	Width	Hole Dia.	Angle
2 1/2"	1/2"	7/8"	45°
2 1/2"	1/2"	7/8"	60°
2 3/4"	1/2"	1"	45°
2 3/4"	1/2"	1"	60"
3"	1/2"	1 1/4"	45°
3"	1/2"	1 1/4"	60°
4"	1/2"	1"	45°
4"	1/2"	1"	60°
4"	1/2"	1 1/4"	45°

Figure 5.76 *Specifications of single angle cutters commercially available.*

Double Angle Cutters

The double angle cutter is a double-sided version of the single angle cutter. They are manufactured with the 45°, 60°, and 90° included angles. This cutter is used to cut a V-groove into a workpiece. By rotating the workpiece 90° you will be able to cut square based pyramids into the face.

Figure 5.77 *A double angle cutter.*

Double Angle Cutters

Diameter	Width	Hole Dia.	Angle
2 3/4"	1/2"	1"	45°
2 3/4"	1/2"	1"	60°
2 3/4"	1/2"	1"	90°
4"	1/2"	1 1/4"	45°
4"	1/2"	1 1/4"	60°
4"	1/2"	1 1/4"	90°
4"	3/4"	1"	45°
4"	3/4"	1"	60°
4"	3/4"	1"	90°
4"	3/4"	1 1/4"	45°
4"	3/4"	1 1/4"	60°
4"	3/4"	1 1/4"	90°
6"	3/4"	1 1/4"	45°
6"	3/4"	1 1/4"	60°
6"	3/4"	1 1/4"	90°
6"	3/4"	1 1/4"	45°
6"	1"	1 1/4"	60°
6"	1"	1 1/4"	90°

Figure 5.78 *Specification of double angle cutters commercially available.*

Convex and Concave Cutters

These two cutters are mirror images of each other – one cuts a radius shape into the workpiece (convex), and the other cuts a radius shape onto the workpiece (concave). This may seem a little confusing, but the convex cutter cuts a half-round or radius shape *into* the workpiece, and the concave cuts a half-round or radius shape *onto* the edge of a workpiece. The use of these shapes may be for decorative and/or structural purposes.

Figure 5.79 *A convex and concave cutter.*

Convex and Concave Cutters

Circle Diameter	Cutter Diameter	Arbor Size
$^{1}/_{8}$"	2"	$^{7}/_{8}$"
$^{1}/_{8}$"	$2^{1}/_{4}$"	1"
$^{3}/_{16}$"	2"	$^{7}/_{8}$"
$^{3}/_{16}$"	$2^{1}/_{4}$"	1"
$^{1}/_{4}$"	2"	$^{7}/_{8}$"
$^{1}/_{4}$"	$2^{1}/_{2}$"	1"
$^{5}/_{16}$"	$2^{3}/_{4}$"	1"
$^{3}/_{8}$"	$2^{1}/_{4}$"	$^{7}/_{8}$"
$^{3}/_{8}$"	$2^{3}/_{4}$"	1"
$^{7}/_{16}$"	3"	1"
$^{1}/_{2}$"	$2^{1}/_{4}$"	$^{7}/_{8}$"
$^{1}/_{2}$"	3"	1"
$^{9}/_{16}$"	3"	1"
$^{5}/_{8}$"	$3^{1}/_{2}$"	$1^{1}/_{4}$"
$^{11}/_{16}$"	3"	1"
$^{3}/_{4}$"	3"	1"
$^{3}/_{4}$"	$3^{1}/_{4}$"	$1^{1}/_{4}$"
$^{13}/_{16}$"	$3^{1}/_{4}$"	1"
$^{7}/_{8}$"	$3^{1}/_{4}$"	1"
$^{7}/_{8}$"	4"	$1^{1}/_{4}$"
$^{15}/_{16}$"	$3^{1}/_{2}$"	1"
1"	$3^{1}/_{4}$"	1"
1"	$4^{1}/_{4}$"	$1^{1}/_{4}$"
$1^{1}/_{16}$"	4"	$1^{1}/_{4}$"
$1^{1}/_{8}$"	4"	$1^{1}/_{4}$"
$1^{1}/_{4}$"	4"	$1^{1}/_{4}$"
$1^{3}/_{8}$"	$4^{1}/_{4}$"	$1^{1}/_{4}$"
$1^{1}/_{2}$"	$4^{1}/_{4}$"	$1^{1}/_{4}$"
$1^{5}/_{8}$"	$5^{1}/_{2}$"	$1^{1}/_{4}$"
$1^{3}/_{4}$"	$5^{1}/_{2}$"	$1^{1}/_{4}$"
$1^{7}/_{8}$"	6"	$1^{1}/_{2}$"
2"	6"	$1^{1}/_{2}$"

Figure 5.80 *A straight tooth Woodruff key cutter.*

Parallel Keyseat Cutters

Parallel keyways are cut with standard size end mills. Keys will be *square* or *rectangular* in shape. The keyway must be machined to within a couple of thousandths of an inch to ensure proper fit. **Note:** If a resharpened end mill is used to cut a keyway, more than one cut may be required to machine the keyway to the proper width. Typical SAE keyway widths include: $^{1}/_{16}$", $^{3}/_{32}$", $^{1}/_{8}$", $^{5}/_{32}$", $^{3}/_{16}$", $^{7}/_{32}$", $^{1}/_{4}$", $^{5}/_{16}$", $^{3}/_{8}$", $^{7}/_{16}$", $^{1}/_{2}$", $^{9}/_{16}$", $^{5}/_{8}$", $^{11}/_{16}$", and $^{3}/_{4}$". ISO keyway widths include: 2 mm, 3 mm, 4 mm, 5 mm, 6 mm, 8 mm, 9 mm, 10 mm, 12 mm, 14 mm, 16 mm, and 18 mm.

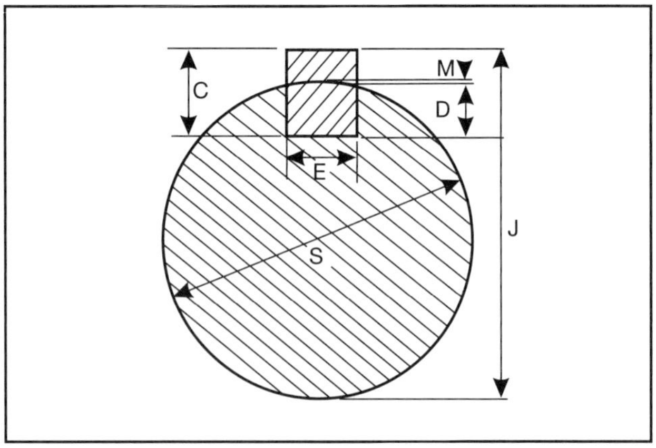

For milling keyseats, the total depth to feed cutter from outside of shaft to bottom of keyseat is M + D, where D is depth of keyseat.
For checking an assembled key and shaft, caliper measurement J between top of key and bottom of shaft is used. J = S - (M + D) + C, where C is depth of key. For Woodruff keys, dimensions C and D can be found in Tables 9 to 11. Assuming shaft diameter S is nominal size, the tolerances on dimension J for Woodruff keys in keyslots are +0.000, -0.010 inch.

Figure 5.81 *This table describes the mathematical formula and suggested key depths for commercially available parallel keyways.*

Woodruff Keyseat Cutters

Woodruff keyseat cutters are used to cut the recess required for a *half moon* key. Cutters will contain several teeth and be manufactured in several diameters and widths. The tooth patterns will be either *straight* or *staggered*.

Tool bits can only be used to cut straight into the workpiece. The tool can be fed in a linear direction parallel to the workpiece long axis if it is used to cut a square keyway.

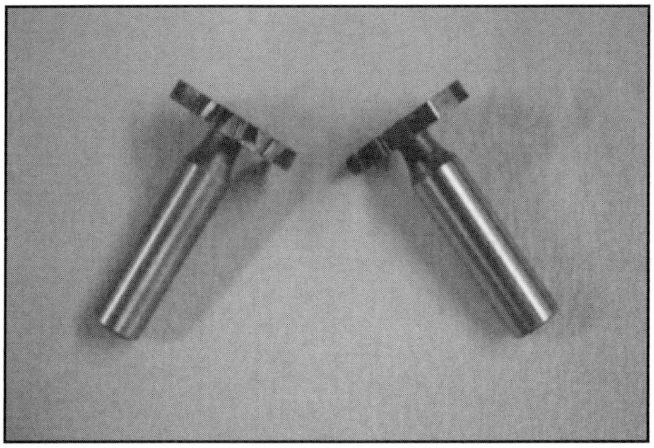

Figure 5.82 Straight tooth Woodruff key cutters.

Tool bits with staggered tooth profiles alternate teeth right and left across the face of the cutter. This type is used for the cutting steel, cast iron, non-ferrous metals, plastic, and non-metals.

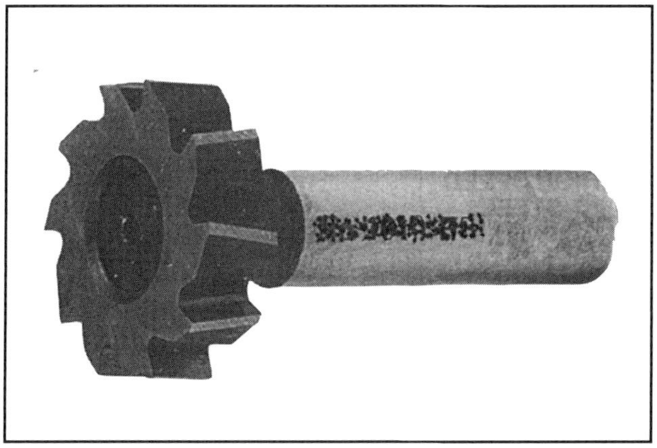

Figure 5.83 A staggered tooth Woodruff key cutter.

Size selection can be somewhat complicated by the different numbering systems (Old Standard and American Standard) used to classify and number each Woodruff key size. Always check the size of the key before you select the tool for your job.

The following chart contains the American Standard numbering system:

Woodruff Keyseat Cutters

Amer. Std. #	Diameter	Face Width
202	$^{1}/_{4}$"	$^{1}/_{16}$"
202-$^{1}/_{2}$	$^{5}/_{16}$"	$^{1}/_{16}$"
302-$^{1}/_{2}$	$^{5}/_{16}$"	$^{3}/_{32}$"
203	$^{3}/_{8}$"	$^{1}/_{16}$"
303	$^{3}/_{8}$"	$^{3}/_{32}$"
403	$^{3}/_{8}$"	$^{1}/_{8}$"
204	$^{1}/_{2}$"	$^{1}/_{16}$"
304	$^{1}/_{2}$"	$^{3}/_{32}$"
305	$^{5}/_{8}$"	$^{3}/_{32}$"
404	$^{1}/_{2}$"	$^{1}/_{8}$"
405	$^{5}/_{8}$"	$^{1}/_{8}$"
406	$^{3}/_{4}$"	$^{1}/_{8}$"
505	$^{5}/_{8}$"	$^{5}/_{32}$"
605	$^{5}/_{8}$"	$^{3}/_{16}$"
506	$^{3}/_{4}$"	$^{5}/_{32}$"
806	$^{3}/_{4}$"	$^{1}/_{4}$"
507	$^{7}/_{8}$"	$^{5}/_{32}$"
606	$^{3}/_{4}$"	$^{3}/_{16}$"
607	$^{7}/_{8}$"	$^{3}/_{16}$"
707	$^{7}/_{8}$"	$^{7}/_{32}$"
608	1"	$^{3}/_{16}$"
708	1"	$^{7}/_{32}$"
1208	1"	$^{3}/_{8}$"
609	$1^{1}/_{8}$"	$^{3}/_{16}$"
807	$^{7}/_{8}$"	$^{1}/_{4}$"
808	1"	$^{1}/_{4}$"
709	$1^{1}/_{8}$"	$^{7}/_{32}$"
809	$1^{1}/_{8}$"	$^{1}/_{4}$"
610	$1^{1}/_{4}$"	$^{3}/_{16}$"
710	$1^{1}/_{4}$"	$^{7}/_{32}$"
810	$1^{1}/_{4}$"	$^{1}/_{4}$"
811	$1^{3}/_{8}$"	$^{1}/_{4}$"
812	$1^{1}/_{2}$"	$^{1}/_{4}$"
1008	1"	$^{5}/_{16}$"
1009	$1^{1}/_{8}$"	$^{5}/_{16}$"
1010	$1^{1}/_{4}$"	$^{5}/_{16}$"
1011	$1^{3}/_{8}$"	$^{5}/_{16}$"
1012	$1^{1}/_{2}$"	$^{5}/_{16}$"
1210	$1^{1}/_{4}$"	$^{3}/_{8}$"
1211	$1^{3}/_{8}$"	$^{3}/_{8}$"
1212	$1^{1}/_{2}$"	$^{3}/_{8}$"

Figure 5.84 *Specifications and application numbers for commercially available Woodruff key cutters.*

Key No.	Nominal Size Key	Keyseat – Shaft					Key Above Shaft		Keyseat – Hub
		Width ***A***	Depth ***B***		Diameter ***F***		Height ***C***	Width ***D***	*Depth* ***E***
		Min.	Max.	+.005 -.000	Min.	Max.	+0.005 -0.005	+0.002 -0.000	+0.005 -0.000
202	1/16 x 1/4	0.0615	0.0630	0.0728	0.250	0.268	0.0312	0.0635	0.0372
202.5	1/16 x 5/16	0.0615	0.0630	0.1038	0.312	0.330	0.0312	0.0635	0.0372
302.5	1/32 x 5/16	0.0928	0.0943	0.0882	0.312	0.330	0.0469	0.0948	0.0529
203	1/16 x 3/8	0.0615	0.0630	0.1358	0.375	0.393	0.0312	0.0635	0.0372
303	3/32 x 3/8	0.0928	0.0943	0.1202	0.375	0.393	0.0469	0.0948	0.0529
403	1/8 x 3/8	0.1240	0.1255	0.1045	0.375	0.393	0.0625	0.1260	0.0685
204	1/16 x ½	0.0615	0.0630	0.1668	0.500	0.518	0.0312	0.0635	0.0372
304	3/32 x ½	0.0928	0.0943	0.1511	0.500	0.518	0.0469	0.0948	0.0529
404	1/8 x ½	0.1240	0.1255	0.1355	0.500	0.518	0.0625	0.1260	0.0685
305	3/32 x 5/8	0.0928	0.0943	0.1981	0.625	0.643	0.0469	0.0948	0.0529
405	1/8 x 5/8	0.1240	0.1255	0.1825	0.625	0.643	0.0625	0.1260	0.0685
505	5/32 x 5/8	0.1553	0.1558	0.1669	0.625	0.643	0.0781	0.1573	0.0841
605	3/16 x 5/8	0.1863	0.1880	0.1513	0.625	0.643	0.0937	0.1885	0.0997
406	1/8 x ¾	0.1240	0.1255	0.2455	0.750	0.768	0.0625	0.1260	0.0685
506	5/32 x ¾	0.1553	0.1558	0.2299	0.750	0.768	0.0781	0.1573	0.0841
606	3/16 x ¾	0.1863	0.1880	0.2143	0.750	0.768	0.0937	0.1885	0.0997
806	¼ x ¾	0.2487	0.2505	0.1830	0.750	0.768	0.1250	0.2510	0.1310
507	5/32 x 7/8	0.1553	0.1558	0.2919	0.875	0.895	0.0781	0.1573	0.0841
607	3/16 x 7/8	0.1863	0.1880	0.2763	0.875	0.895	0.0937	0.1885	0.0997
707	7/32 x 7/8	0.2175	0.2193	0.2607	0.875	0.895	0.1093	0.2198	0.1153
807	¼ x 7/8	0.2487	0.2505	0.2450	0.875	0.895	0.1250	0.2510	0.1310
608	3/16 x 1	0.1863	0.1880	0.3393	1.000	1.020	0.0937	0.1885	0.0997
708	7/32 x 1	0.2175	0.2193	0.3237	1.000	1.020	0.1093	0.2198	0.1153
808	¼ x 1	0.2487	0.2505	0.3080	1.000	1.020	0.1250	0.2510	0.1310
1008	5/16 x 1	0.3111	0.3130	0.2768	1.000	1.020	0.1562	0.3135	0.1622
1208	3/8 x 1	0.3735	0.3755	0.2455	1.000	1.020	0.1875	0.3760	0.1935

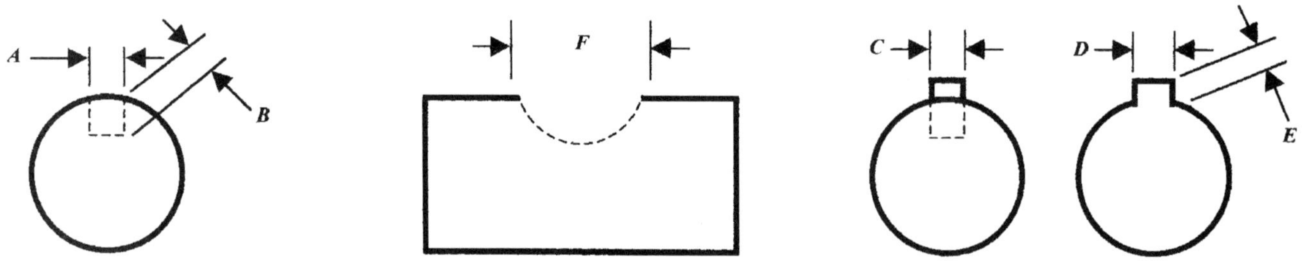

Figure 5.85 *This table describes the suggested key depths for several commercially available Woodruff key cutters.*

Involute Gear Cutters

Involute gear cutters are used to mill *spur* gears having a $14^{1}/_{2}°$ or 20° pressure angle. This group of cutters conform to recognized gear tooth profiles with diametrical pitches from 2 to 48, and cut a range of teeth from 12 to rack with eight specially ground cutters (per each diametrical pitch).

Figure 5.86 *An involute gear cutter.*

Involute Gear Cutters

Cutter Number	Range of Teeth Cut	Cutter Number	Range of Teeth Cut
1	135 to rack	5	21 to 25
2	55 to 134	6	17 to 20
3	35 to 54	7	14 to 16
4	26 to 34	8	12 to 13

Figure 5.87 *The cutter blade number (1, 2, 3, 4, 5, 6, 7, or 8) is specific to the number of teeth you wish to cut.*

When ordering gear cutters, be sure to specify the pressure angle ($14^{1}/_{2}°$ or 20°), number of teeth to be cut, and the arbor hole size. Cutters are typically made from high-speed steel, with or without coatings. Gear cutters must be mounted onto a stub shaft, or used with the right angle adapter.

Figure 5.88 *This gear blank is ready to cut.*

Gear blanks are mounted onto an arbor and positioned (in relationship to the gear tooth location) with an *index (dividing) head.*

Specialty End Mills

What do you do when a part must have a special shape or form machined onto the surface? Have a special end mill made. The majority of special end mills are made from existing types. They are reground and shaped to fit a particular purpose. Most cutter resharpening services can perform this type of modification.

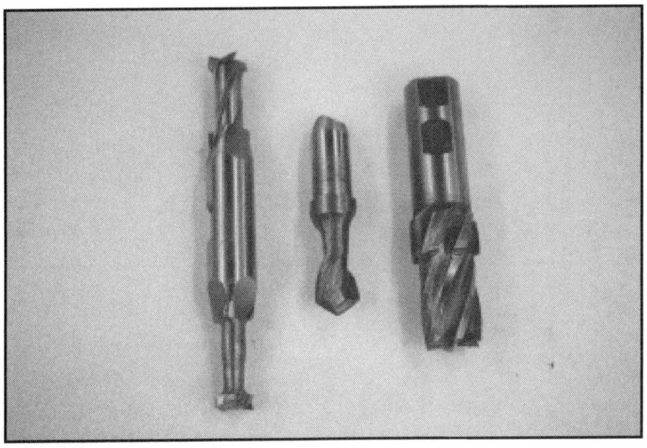

Figure 5.89 *Specialty end mills.*

Carbide Insert Types

One of the most significant improvements in cutter technology is the use of replaceable carbide inserts. High-speed steel and solid carbide end mills cannot keep pace with the insert-type counterparts under the strains of high production demands. The majority of insert-type cutters use a steel body onto which the insert is attached. Most cutters may have from one to six inserts. Large diameter flycutters have even more.

You need to understand what the inserts are and the different types and configurations commercially available. As previously discussed, inserts come in various grades or alloy compositions, e.g., C-2, C-5, C-6, etc. These types can be coated to extend tool life expectancy.

To increase the production rate of simple carbide, many manufacturers have developed other materials for use as inserts. Namely, *cubic boron nitride* (CBN) and *poly crystalline diamond* (PCD). In the majority of insert applications, the CBN or PCD is applied to one corner or edge of a carbide substrate material. This one edge can outperform several carbide inserts. Each material has specific applications. CBN inserts are used primarily on steel and cast iron, PCD inserts give their best performance when used with light non-magnetic alloys.

Inserts come in several different shapes and thicknesses. They use a descriptive set of terms, prefixes, and suffixes to describe them. The *inscribed circle* (I.C.) describes the diameter of the insert when all sides are tangent to the outside edge of a circle. *Thickness* (T) describes how thick the insert is. Instead of a sharp point at each corner, most inserts use a *radius* (R). The amount of radius (which is standardized) depends upon the inscribed circle of the insert, but sizes of .008", $^1/_{64}$", $^1/_{32}$", $^3/_{64}$", and $^1/_{16}$" are common.

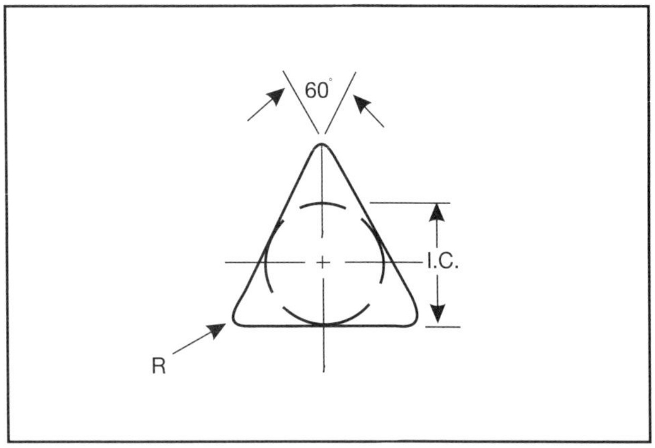

Figure 5.90 *The inscribed circle is used to denote insert size.*

Inserts are made with a *positive* or *negative* rake. Positive rake inserts can be used in tools with a ZERO degree cutting angle location in relationship to the workpiece. These inserts have a *relief angle* ground onto the insert. Negative rake inserts must be placed at a minus angle cutting angle in relationship to the workpiece. Never interchange positive and negative rake inserts with one another.

An innovation to insert design is the *chip control* groove. This groove may run parallel to the insert cutting edge, or it may have a *bumpy* surface texture to increase the flow of coolants to the cutting area.

Figure 5.91 *This selection of inserts includes both "plain" and "bumpy" surfaces.*

The following chart illustrates some of the different shapes and their attributes:

Insert Types (Typical)

Insert Shape	Description
TPG	Triangle Positive Rake
TPU	Triangle Positive Rake (Utility)
TNMA	Triangle Negative Rake
TNMG	Triangle Negative Rake w/Chip Control
TPMR	Triangle Positive Rake w/Chip Control
SPG	Square Positive Rake
SNG	Square Negative Rake
SNU	Square Negative Rake (Utility)
SNMA	Square Negative Rake Insert
RNG	Round Negative Rake
CNMG	80° Diamond Negative Rake
CNMP	80° Diamond Positive/Negative Rake
DNMG	55° Diamond Negative Rake
VNMG	35° Diamond w/Hole & Molded Chip Breaker

Figure 5.92 *These are but a small sample of various inserts commercially available.*

Boring Tools

Boring heads allow the machine operator the opportunity to machine holes with unusual sizes. They use a replaceable high-speed steel or carbide-tipped tool bit, and come in a wide range of shank diameters, lengths, and minimum bore sizes. Both tip types allow for customization for special applications. For best results, keep the length of the boring tool as short as possible, as this will increase rigidity.

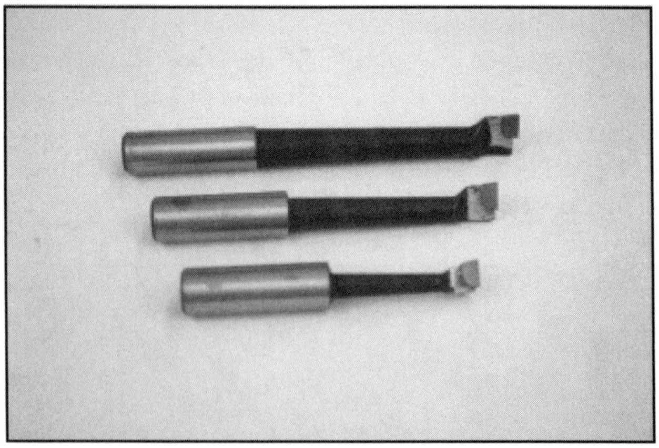

Figure 5.93 *Adjustable boring heads can use a wide variety of tool lengths and diameters.*

Boring Tools

Shank Diameter	Max. Bore Depth	Min. Bore Size
3/8"	5/16"	1/16"
"	11/32"	3/32"
"	1/2"	1/8"
"	13/16"	3/16"
"	1 1/8"	1/4"
"	1 1/2"	5/16"
"	1 3/4"	3/8"
1/2"	5/16"	1/16"
"	11/32"	3/32"
"	1/2"	1/8"
"	13/16"	3/16"
"	1 1/8"	1/4"
"	1 1/2"	5/16"
"	1 3/4"	3/8"
"	2 3/16"	1/2"

Figure 5.94 *Specifications of boring tools commercially available.*

Tapping Holes With Hand Tools

There are two widely used methods to tap holes into a workpiece on the mill: with a tap handle and guide, or with the machine quill. Let's take a look at both.

Using a hand tap with a tap handle is a basic machinist bench operation. A tap is placed in the handle, the tap is aligned with the hole, then the tap is twisted into the hole to cut the threads. Easy enough – but the tap does require a bit more alignment to get it correctly aligned with the hole. Using a tap guide while the part is still set-up in the mill will ensure that the tap enters the hole at the same exact angle that the hole was drilled.

Tap guides are typically 1/2" (12.7 mm) in diameter and will fit neatly in a drill chuck or collet. The pin used to guide the tap into the hole has two pointed ends – one with a standard taper, the other a reverse taper. The standard taper is used for taps with a center hole drilled in the drive end. The reverse taper end is used on small diameter taps that have a taper on the end of the tap. This pointy end of the tap will slip right into the reverse taper of the guide pin.

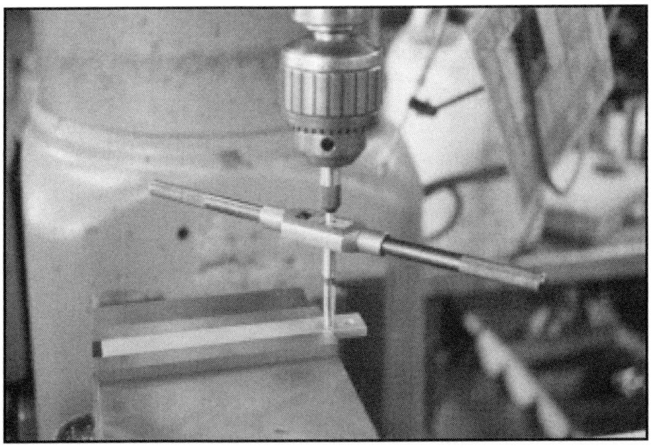

Figure 5.95 *This tap guide is used to help align and start a hand tap into a predrilled hole using the quill of the mill.*

To use the tap guide, install the guide pin to match the tap being used as described above. Install the unit into the chuck or collet. Place the tap (with the tap handle installed) into the predrilled hole. Lower the tap guide until the guide pin is aligned with the tap, then, compress the quill another .400"–.500" (10.0–13.0 mm). This will cause the pin to retract into the tap guide assembly. Start turning the tap and be sure to apply the appropriate lubricant. As the tap enters the hole, the guide pin will follow the tap because the pin is spring loaded. Once you have at least 4–6 threads cut, remove the tap guide and continue the tapping process by hand.

The next method used to tap holes requires a considerable amount of experience on the part of the machinist. The process involves using the mill motor to drive the tap, and the machinist pulling and pushing the quill to feed the tap into and out of the workpiece. This process is best performed on parts with *through holes*, and not ones with *blind holes*. As the terminology intimates, a through hole is one that passes all the way through the workpiece with no obstruction. A blind hole is closed off at the end and will not allow the drill or tap to pass all the way through.

Figure 5.96 *This hole is being tapped with the mill using the Forward/Reverse switch to control tap direction.*

Keep in mind that the tap has a certain number of threads per inch (threads per millimeter). Movement of the quill will need to mimic this thread pitch, both in and out of the hole. Move the quill too fast, or too slow, and the tap will typically break-off in the workpiece. An ugly thought but one that must be remembered when attempting to speed things up. Another problem when moving the quill too fast or slow may be an increase in the width of the threads. This will naturally cause the threaded fastener to be very loose fitting. The threads will eventually fail in the workpiece because they are too thin to hold the fastener properly.

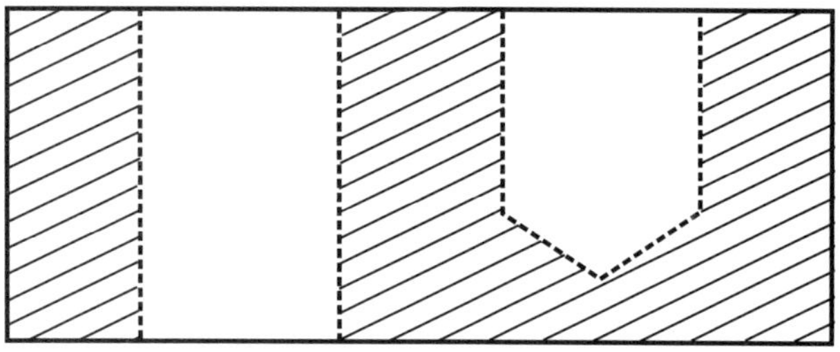

Figure 5.97 Examples of a through hole (left) and blind hole (right).

Tap Drills

Before a location can be threaded with a tap, a *tap drill* must be used to drill the hole. Tap drills are smaller than the *body* (diameter) size of the fastener. Remember, a drill bit is considered a roughing tool. If the bit is not sharp or the cutting lips are not equal, an oversize hole will result.

When discussing internal threading, we want to look at the *percentage* of thread contact with the fastener (screw or bolt). In general, a 55% (.5) –60% (.6) is typical, but 75% (.75) contact area is used to provide an extra margin of safety. The majority of commercially available taps will cut threads with a 75% contact area. This percentage can be *adjusted* by increasing or decreasing the tap drill diameter.

In general, 55% (.55) contact area is used when the engagement length is at least 1.5:1 times the thread diameter. Softer ductile materials may allow the use of a slightly larger tap drill than brittle materials, e.g., gray cast iron. The following chart will provide you with the universally accepted tap drill sizes used for tapping standard holes.

Tap Drill Size Chart

Thread Size	Tap Drill
$^1/_4$-20	#7
$^1/_4$-28	#3
$^5/_{16}$-18	F
$^5/_{16}$-24	I
$^3/_8$-16	$^5/_{16}$"
$^3/_8$-24	Q
$^7/_{16}$-14	U
$^7/_{16}$-20	$^{25}/_{64}$"
$^1/_2$-13	$^{27}/_{64}$"
$^1/_2$-20	$^{29}/_{64}$"

Figure 5.98 Use this chart to determine what drill to use when tapping holes.

Large Hole Boring Tools

This group of cutting tools use a special tool holder and blade assembly. A 1" (25.4 mm) diameter pilot hole must be bored into the workpiece before the tool can be used. The pilot hole creates a pathway used to support the cutter assembly. Blades from $1^1/_8$" to 4" are available in $^1/_{16}$" incremental sizes up to 3", and $^1/_2$" increments up to 4". This tool should be considered when machining large diameter holes rather than using a boring head to remove small amounts of stock.

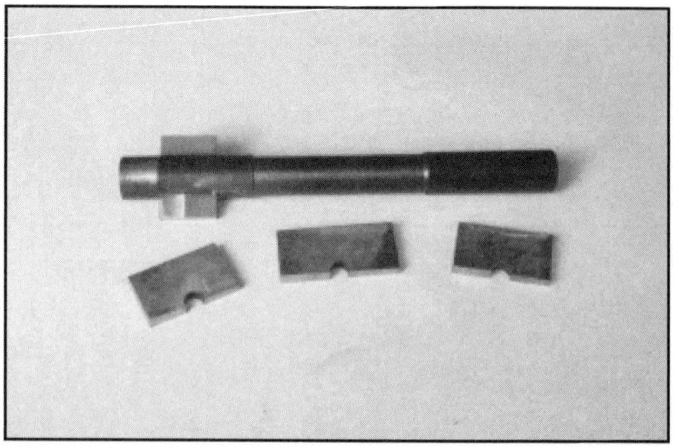

Figure 5.99 This is a large hole boring tool and a selection of different size interchangeable cutter blades.

Large Hole Cutter Blade Diameter Chart

Blade Diameter	Blade Diameter
$1^1/_8$"	$2^1/_8$"
$1^5/_{32}$"	$2^3/_{16}$"
$1^3/_{16}$"	$2^1/_4$"
$1^7/_{32}$"	$2^5/_{16}$"
$1^1/_4$"	$2^3/_8$"
$1^5/_{16}$"	$2^7/_{16}$"
$1^3/_8$"	$2^1/_2$"
$1^7/_{16}$"	$2^9/_{16}$"
$1^1/_2$"	$2^5/_8$"
$1^9/_{16}$"	$2^{11}/_{16}$"
$1^5/_8$"	$2^3/_4$"
$1^{11}/_{16}$"	$2^{13}/_{16}$"
$1^3/_4$"	$2^7/_8$"
$1^{13}/_{16}$"	$2^{15}/_{16}$"
$1^7/_8$"	3"
$1^{15}/_{16}$"	$3^1/_2$"
2"	4"
$2^1/_{16}$"	

Figure 5.100 Large hole boring tool blades are available in several sizes.

Counterbore Tools

Counterbore tools are used to enlarge a previously drilled hole. They are commonly used to recess hex-head cap screws to give a part a more *finished* appearance. There are two types of counter bore tools: those with interchangeable pilots and those with standard size integral pilots. The integral piloted tools are typically made just for counterboring socket head cap screws. The interchangeable pilot types are used where specific size holes and counterbore sizes are necessary. Commercially available pilots and counterbores are available in 1/32" and 1/16" diameter size increments.

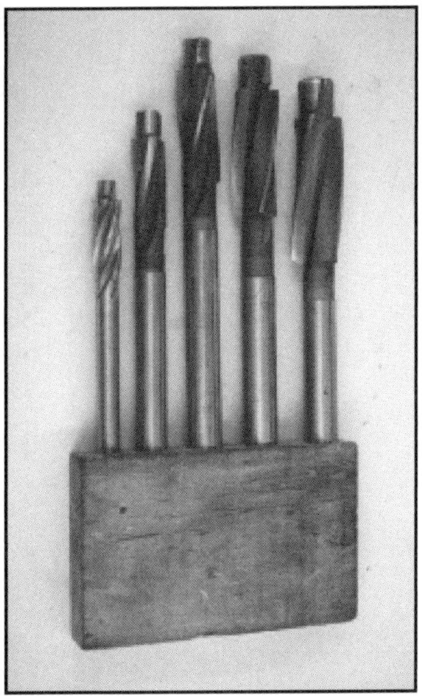

Figure 5.101 *This is a set of socket head cap screw counterbores.*

Cap Screw Counterbore Chart

Screw Size	Pilot Size	Cutter	Shank Diameter	OAL
#5	.141"	.221"	3/16"	3"
#6	.150"	.242"	7/32"	3"
#8	11/64"	19/64"	1/4"	3"
#10	13/64"	21/64"	5/16"	3 1/2"
1/4"	9/32"	13/32"	3/8"	5"
5/16"	11/32"	1/2"	7/16"	5"
3/8"	13/32"	19/32"	1/2"	6"
7/16"	15/32"	11/16"	1/2"	7"
1/2"	17/32"	25/32"	1/2"	7 1/2"
5/8"	21/32"	31/32"	3/4"	7 1/2"
3/4"	13/16"	1 3/16"	1"	8"
7/8"	15/16"	1 3/8"	1"	8"
1"	1 1/16"	1 9/16"	1"	10"

Figure 5.102 *Standard cap screw counterbore sizes.*

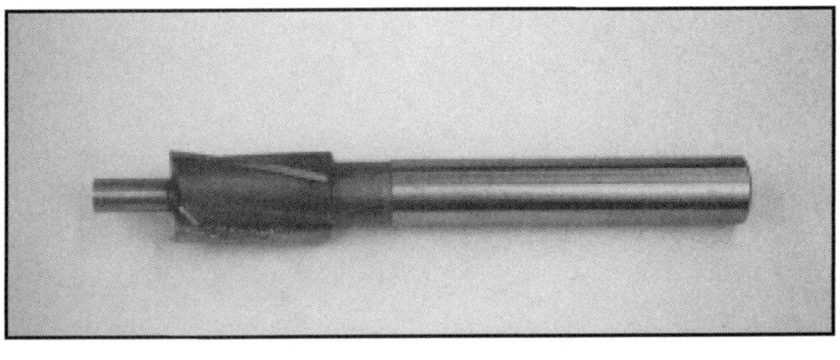

Figure 5.103 *This is a counterbore with an interchangeable pilot.*

The pilots used for this group of counterbores are supplied with various shank diameters to fit the hole in the middle of the cutter. Shank size is dependent upon the diameter of the counterbore. The head size of the pilot can be purchased in 1/32" size increments, or made from bar stock to fit any size desired, as long as the shank diameter will fit the counterbore hole.

Counterbore Size Chart

Diam.	OAL	Shank Size & Length	Number of Flutes	Hole Diam.
3/16"	3"	15/64 × 2 1/8"	3	3/32"
7/32"	3"	15/64 × 2 1/8"	3	3/32"
1/4"	3 13/16"	15/64 × 3 1/16"	3	3/32"
9/32"	3 13/16"	17/64 × 3 1/6"	3	3/32"
5/16"	3 13/16"	19/64 × 3 1/6"	3	3/32"
11/32"	3 13/16"	5/16 × 3 1/16"	3	3/32"
3/8"	4 1/16"	5/16 × 3 1/16"	3	5/32"
13/32"	4 1/16"	3/8 × 3 1/16"	3	5/32"
7/16"	4 1/16"	3/8 × 3 1/16"	3	5/32"
15/32"	4 5/16"	7/16 × 3 1/16"	3	3/16"
1/2"	4 5/16"	7/16 × 3 1/16"	3	3/16"
17/32"	4 5/16"	1/2 × 3 1/16"	3	3/16"
9/16"	4 5/16"	1/2 × 3 1/16"	3	3/16"
19/32"	5 1/8"	1/2 × 3 7/8"	3	3/16"
5/8"	5 1/8"	1/2 × 3 7/8"	3	3/16"
21/32"	5 1/8"	1/2 × 3 7/8"	3	3/16"
11/16"	5 1/8"	1/2 × 3 7/8"	3	3/16"
23/32"	5 3/8"	1/2 × 3 7/8"	3	1/4"
3/4"	5 3/8"	1/2 × 3 7/8"	3	1/4"
25/32"	5 3/8"	5/8 × 3 7/8"	3	1/4"
13/16"	5 3/8"	5/8 × 3 7/8"	3	1/4"
27/32"	5 3/8"	5/8 × 3 7/8"	3	1/4"
7/8"	5 3/8"	3/4 × 3 7/8"	3	1/4"
29/32"	6 1/8"	3/4 × 3 7/8"	3	1/4"
15/16"	6 1/8"	3/4 × 4 5/8"	3	1/4"
31/32"	6 3/8"	3/4 × 4 5/8"	3	5/16"
1"	6 3/8"	3/4 × 4 5/8"	3	5/16"
1 1/16"	6 3/8"	3/4 × 4 5/8"	3	5/16"
1 1/8"	6 3/8"	1 × 4 5/8"	3	5/16"
1 3/16"	6 3/8"	1 × 4-5/8"	3	5/16"
1 1/4"	6 5/8"	1 × 4-5/8"	5	3/8"
1 5/16"	6 5/8"	1 × 4-5/8"	5	3/8"
1 3/8"	6 5/8"	1 × 4 5/8"	5	3/8"
1 7/16"	7 7/8"	1 1/4 × 5 7/8"	5	3/8"
1 1/2"	7 7/8"	1 1/4 × 5 7/8"	5	3/8"
1 5/8"	8 1/8"	1 1/4 × 5 7/8"	5	7/16"
1 11/16"	8 1/8"	1 1/4 × 5 7/8"	5	7/16"
1 3/4"	8 1/8"	1 1/4 × 5 7/8"	5	7/16"
1 13/16"	8 1/8"	1 1/2 × 5 7/8"	5	7/16"

Figure 5.104 *Counterbore with an interchangeable pilot sizes.*

Commercially manufactured counterbore diameters larger than $1^{13}/_{16}$" are also available; check your tool supplier for sizes and recommendations.

Some Further Thoughts on Cutting Tools . . .

There is a relationship between cutter materials and machine stock, and whether or not the cut is continuous or interrupted. Aluminum is one such case. Unlike cast iron, aluminum requires the usage of coolants. Cast iron has a high percentage of graphite within its matrix. The graphite works and acts like a natural lubricant. Aluminum tends to be somewhat *sticky* when machined. Small *whiskers* will attach themselves to the cutter (sometimes called edge loading). Left alone, the whiskers will cause the workpiece to be cut undersize and impart a rough surface finish. Aluminum responds well to PCD tooling, especially when interrupted flycutting is performed. The diamond coating inhibits edge loading, unlike high-speed or carbide.

Figure 5.105 *This cylinder head is a typical example of an interrupted cut. The flycutter has a PVD insert cutter bit installed.*

Again, it cannot be overstated that you should call your cutting tool supplier for their recommendation when considering the purchase of any new cutting tool(s). They have spent many thousands of man-hours and dollars to research the best cutter-material combinations for their products. Unless you have an advanced degree in metallurgy and fully understand the characteristics and relationships between cutting tools and the materials that they cut, stick to the manufacturer's specifications and recommendations. This same thinking must be applied to tool holders. The negative and positive cutter positions have been carefully thought out to ensure proper material removal rates.

One final word about safety . . . Think. Cutters, by and large, are sharp and require your full attention when installing or removing them. Use appropriate hand and eye protection when handling or working around any cutting tool.

NOTE: All dimensions for cutters listed in this chapter are approximate. Actual sizes may vary by manufacturer. Check with your tool supplier to obtain exact dimensions and performance data.

Next . . .

The next chapter contains information about work holding tools. If the workpiece is not properly held, the whole machining process is for nothing. You will learn about vises and other devices used to clamp, secure, or hold the workpiece for machining.

Notes:

Chapter 5 ◆ Study Questions

1. Machinist A says that tool coatings such as TiN are applied to tools to increase their resistance to shattering, Machinist B says that vapor deposition-type coatings are used to increase wear resistance. Who is correct?
   A. Machinist A
   B. Machinist B
   C. Both A and B
   D. Neither A nor B

2. Machinist A says that center-cutting end mills can be used to bore a hole, Machinist B says that some end mills with center-holes in them cannot be used to bore holes. Who is correct?
   A. Machinist A
   B. Machinist B
   C. Both A and B
   D. Neither A nor B

3. Machinist A says that the amount of taper specified on a taper end mill indicates the amount of taper of the included angle, Machinist B says that T-slot cutters are also used to cut Woodruff-key slots. Who is correct?
   A. Machinist A
   B. Machinist B
   C. Both A and B
   D. Neither A nor B

4. Machinist A says that ball-type end mill diameters are calculated from the center of the tool bit, not the diameter, Machinist B says that straight keyways can also be cut with an end mill. Who is correct?
   A. Machinist A
   B. Machinist B
   C. Both A and B
   D. Neither A nor B

5. Machinist A says that through-holes are those holes where the hole passes completely through the workpiece, Machinist B says that multi-flute end mills can be run faster than two-flute types. Who is correct?
   A. Machinist A
   B. Machinist B
   C. Both A and B
   D. Neither A nor B

6. Machinist A says that PCD tools are made from cubic boron nitride, Machinist B says that diamond plated tools work extremely well in light alloy materials. Who is correct?
   A. Machinist A
   B. Machinist B
   C. Both A and B
   D. Neither A nor B

7. Machinist A says that carbide is manufactured in several grades, Machinist B says that graded carbide must be chosen with material type and cutting style in mind. Who is correct?
   A. Machinist A
   B. Machinist B
   C. Both A and B
   D. Neither A nor B

8. Machinist A says that plain mill cutters are typically used on horizontal milling machines, Machinist B says that center-drills are used to locate a drilled hole location. Who is correct?
   A. Machinist A
   B. Machinist B
   C. Both A and B
   D. Neither A nor B

9. Machinist A says that some mill cutters may have brazed carbide cutting edges, Machinist B says that most end mills are made from solid stock. Who is correct?
   A. Machinist A
   B. Machinist B
   C. Both A and B
   D. Neither A nor B

Calculate the spindle rpm for the following cutter sizes:

10. .750" diameter, 125 sfm = rpm
11. .138" diameter, 50 sfm = rpm
12. .518" diameter, 300 sfm = rpm
13. .375" diameter, 1.5 ipm = rpm
14. .875" diameter, 4 ipm = rpm
15. 5.000" diameter, 1.75 ipm = rpm
16. 3.500" diameter, 90 sfm = rpm
17. .0625" diameter, 140 sfm = rpm
18. .750" diameter, 1.5 ipm = rpm
19. .4375" diameter, 2 ipm = rpm
20. .625" diameter, 600 sfm = rpm
21. 12" diameter, 45 sfm = rpm
22. 47.500 diameter, 70 sfm = rpm

Chapter 6
Work Holding Tools

Workpieces must be held securely to the mill table for operator and machine safety. To that end, you must learn how to use a variety of tooling to position and secure the workpiece. Parts must be held in such a way that they will not move or vibrate during machining.

There are a variety of tools used to hold parts. Some are as simple as clamps and vises, while others are far more sophisticated like tilting tables and index (dividing) heads. Your understanding and care of work holding tools will serve you very well for a long time.

Table T-Slots

The primary method used to secure tooling to the mill table is *T-slot hold-down*. These slots are milled into the table when it is manufactured. The nominal slot width is $^5/_8$". The slots are machined from one end of the table to the other. The majority of mills have at least three T-slots. One in the middle of the table and two flanking slots.

Figure 6.1 *T-slots are machined into the table. They are used in just about every tool or workpiece mounting.*

In order to secure the workpiece or tooling to the table, a *T-slot nut* or *T-bolt* is used. They are made to slide freely in the T-slot. The T-slot nut has a threaded hole into which bolts or studs can be screwed. The most common thread size is $^1/_2$"-13. T-bolts are made with $^1/_2$" shanks and come in various lengths.

Kits containing a variety of T-slot nuts, studs, and coupling nuts, or T-slot bolts are available through machine distributors and shop supply houses. The majority of these kits also include a set of clamps, studs, flanged nuts, coupling nuts and step-blocks for clamping purposes.

Figure 6.2 *A set of T-slot bolts and clamps typically used to hold or clamp material and tooling onto the machine.*

T-slot bolts, while not as common as T-slot nuts, do provide an excellent means to clamp parts or tooling to the mill table. They are routinely used to clamp tooling like vises and rotary tables in place. They require the use of a *nut and washer* or *flanged nut* to tighten the tooling to the table.

Thread Size	Thread OAL
$^1/_2$"-13	1$^1/_2$"
$^1/_2$"-13	2"
$^1/_2$"-13	2$^1/_2$"
$^1/_2$"-13	3"
$^1/_2$"-13	3$^1/_2$"
$^1/_2$"-13	4"
$^1/_2$"-13	4$^1/_2$"
$^1/_2$"-13	5"
$^1/_2$"-13	5$^1/_2$"
$^1/_2$"-13	6"
$^1/_2$"-13	7"
$^1/_2$"-13	8"
$^1/_2$"-13	9"
$^1/_2$"-13	10"
$^1/_2$"-13	12"

Figure 6.3 *Specifications for commercially available T-slot bolts.*

There are smaller and larger T-bolts available for smaller and larger milling machines, including: $^3/_8$"-16, $^5/_8$"-11, $^3/_4$"-10, $^7/_8$"-9, and 1"-8 which all come in a variety of lengths.

T-slot nuts and T-bolts use the same body dimensions at the nut end, $^1/_2$"-13. Some special T-slot nuts have both $^3/_8$"-16 and $^1/_2$"-13 threads cut into them. This allows the use of smaller and/or larger studs to clamp various sized workpieces. Very small parts, in general, cannot be clamped with standard $^1/_2$" types.

Because of the $^{1}/_{2}$"-13 threads we can use studs or bolts to couple the workpiece or tool to the T-slot nut. Studs come in different lengths. They are threaded at both ends, or from end to end (all thread rods). *All thread* rods can be cut to length to suit almost any application because they are available in up to 36" lengths. Double-ended studs come in the following lengths: 3", 4", 5", 6", 7", and 8". If the 8" is still too short, coupling nuts can be used to increase the length.

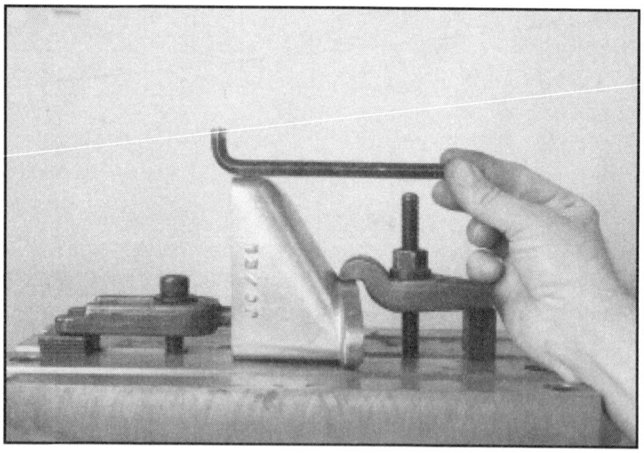

Figure 6.4 *T-slot bolts and clamps are used to hold a piece of stock onto the mill table. Always make sure that the bolts are located below the top of the workpiece.*

The use of $^{1}/_{2}$"-13 cap screws is also an acceptable alternative to studs. However, many machinists prefer the use of socket-head cap screws instead of hex-head cap screws. Socket-head cap screws have a minimum bolt rating of Grade 8.

Figure 6.5 *Grade 8 fasteners should be used to clamp the workpiece to the mill table. Socket-head cap screws are used by many machinists to hold parts to the mill table.*

Nuts used with $^{1}/_{2}$" studs will typically have a larger nut size ($^{7}/_{8}$" across the flats) than standard SAE and USS types ($^{3}/_{4}$" across the flats). Nuts should be used with a heavy-duty washer in order to spread the clamping load onto the clamp or tool. *Flanged nuts* combine both nut and washer into one unit and are the preferred choice to clamp parts to the machine.

Figure 6.6 *Flanged nuts eliminate the use of washers.*

In those special circumstances where a flat washer cannot be used where a workpiece or tool must be clamped slightly off center, there are *spherical washers*. These washers have a top and bottom half — one with a slight concave shape, the other is slightly convex. This allows for a slight degree of bolt misalignment with the clamp.

Figure 6.7 *Spherical washers allow a slight degree of bolt misalignment with the clamp.*

Coupling nuts are the same size (across the flats) as clamp nuts, however, they are three to four times longer. They allow two studs to be connected end to end in order to achieve the desired bolt length necessary for clamping purposes. Coupling nuts can also be used with a standard bolt to create a leveling or support jack for the workpiece. This can be very handy when leveling bulky or awkward parts to the machine table.

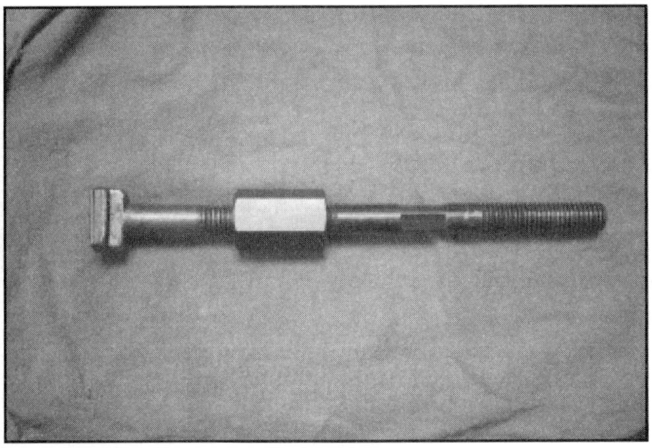

Figure 6.8 *Coupling nuts can be used to extend the range or reach of standard studs.*

Step blocks are used to support one end of the clamp. They come in a variety of sizes and widths to suit most situations. A series of *stair-steps* are milled into the long side of the block. They allow you to raise and lower the block height within a predefined range. This is necessary in order to position the clamp correctly in relationship to the workpiece or tool height. These blocks are not manufactured to high precision tolerances; do not confuse them with parallels.

Figure 6.9 *Step blocks allow for a wide range of support for clamps. They can be stacked to increase their effective work height.*

Clamps come in a wide variety of sizes. The most common include: U-clamp, gooseneck, finger, and plain universal strap clamps. They are drop-forged for strength and durability. Universal adjustable clamp sets are designed for holding parts within a predefined height limit without the need for step blocks. Leveling clamps are used for holding parts with a flat bottom side, or for plate stock. They have three or four steps that locate the workpiece above the top of the table and allow complete tool penetration without causing damage to the machine table. Edge clamps hold a workpiece from the side or edge. Rather than clamping *down*, the workpiece is held in place with *side pressure*. Cam action clamps use an eccentric rotary motion to side-clamp a workpiece when the clamp is activated. Step clamps when used with step blocks provide a very stable clamping mechanism.

Figure 6.10 *A variety of forged hold-down clamps.*

Proper Clamping

The importance of proper clamping is another area that cannot be overstressed. Clamps should be placed so that they are as parallel as possible to the machine table. This means that the clamping *down-force* of the workpiece should be applied straight down and through the part.

Never position a clamp at an angle and then proceed to tighten the clamp bolts. As previously discussed, spherical washers can be used for those odd situations where the clamp must be positioned at other than ideal locations. **Note:** Use your own best judgment when applying pressure at less than ideal situations. The typical problem when applying pressure at other than right angles is that the workpiece will move out of position. This is a real consideration if you have already indicated the part for set-up location. Realistically, the workpiece should be indicated after it is clamped.

Figure 6.11 *This illustrates the proper alignment between the clamp and machine table.*

Figure 6.12 This illustrates incorrect clamping, the step block is too low and will cause the part to "push" out of position.

Figure 6.13 This also illustrates incorrect clamping, the step block is too high (note the extreme angle of bolt) and will cause the clamp to "dig" into the surface of the part.

Parallels

There will be times when a workpiece needs to be raised above the surface of the table. To do this you will need a pair of *parallels*. Parallels are made from hardened steel and are ground in paired sets. They are available in various widths and heights to suit most applications.

Short parallels can also be used to support work held in a vise. Or they can be placed outboard (on both sides of the vise) to support longer workpieces. Two or three sets of parallels of various heights will be very handy for set-up purposes.

Vises

A *vise* will be your most commonly used accessory for holding workpieces. They are rugged tools made from cast iron. Vises have one fixed (non-movable) and movable jaw (one that moves). The base of the vise is made so that it will align the jaws parallel or perpendicular to the table T-slots, or with a rotary feature that will allow the vise to be rotated off axis. *Angle* and *compound angle* vises have a provision for placing the workpiece at a single angle (angle vises) or at two angles at one time (compound angle).

Mill-Type Vise

This group of vises is different than those used throughout the machine shop. They are built for use on a milling machine. The base of a mill vise contains slots that enable it to be clamped to the mill table.

Figure 6.14 A typical mill vise.

Mill vises are available in several designs, including straight lock, angle lock, and those with rotary or swivel bases. Straight jaw vises are the most common. They clamp the workpiece between two jaws as a lead screw is tightened. The angle lock vise is more effective at holding a workpiece. As the clamping pressure is increased, so is the downward pressure. For every pound of forward clamping force, one-half pound downward force is also generated. This effect enhances the clamping action dramatically.

Vises with rotary bases allow you to position a vise at different angles to the X- and Y-axes, and makes angle cutting much easier.

Vise jaws are a component part of any vise. They are typically made of hardened steel, tempered aluminum, or soft brass

Figure 6.15 Most mill vises are equipped with smooth jaws.

depending upon the material being clamped. Plastic jaws can also be used for special clamping situations.

For most clamping applications, vise jaws are flat and smooth. Occasionally, however, there is a need for *serrated jaws* in certain clamping situations. Serrated jaws must be used with care because they will leave tell-tale clamp marks on the surface of the workpiece. This can be especially bad if the workpiece must maintain a specific surface finish. Plus, the indentations left in the workpiece may cause *stress risers*, a condition that can lead to component failure.

For holding small-diameter round workpieces, use a pair of vise jaws with V-grooves. This type of jaw will normally have two grooves – one parallel to the worktable, and one perpendicular (90°) to the table. These grooves do not supply sufficient clamping area for workpieces much larger than $^3/_4$" (19 mm) diameter unless specially modified.

Figure 6.16 *V-groove jaws are used to hold workpieces parallel or at right angles to the worktable.*

Soft jaws are very handy because they can be machined to hold one or several workpieces at one time.

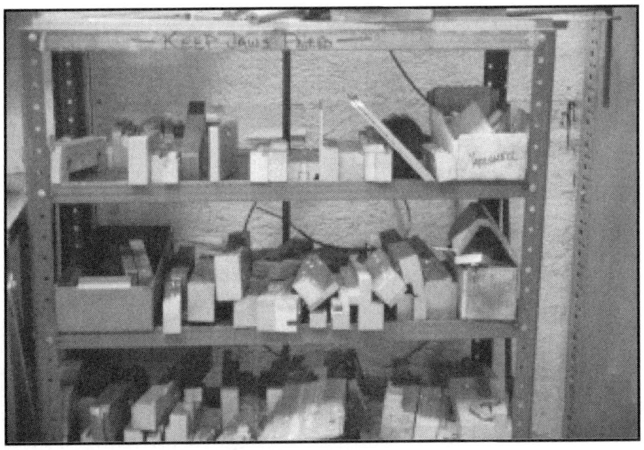

Figure 6.17 *Soft jaws are required for some material types. This shelf unit is holding soft jaws used for different jobs; they are always stored in pairs.*

Don't *overtighten the vise. This cannot be stressed enough. Overtightening the vise can distort your workpiece to a point where the piece must be scrapped. Easy does it!*

Angle-Type

When the workpiece requires machining at a specific angle, use an angle-type vise. The base of the vise should be bolted securely to the machine table, the upper portion is hinged for angle operation. A pair of adjustable supports are used to stabilize the vise at its prescribed angle. The vise can be adjusted from 0° to 90°.

Figure 6.18 *An angle-type vise.*

Compound Angle-Type

Compound angle vises are used for those multiple angle machining operations. The base of the compound angle vise is bolted securely to the machine table, just like the angle vise. The vise has provisions for two individual angle locations on the workpiece.

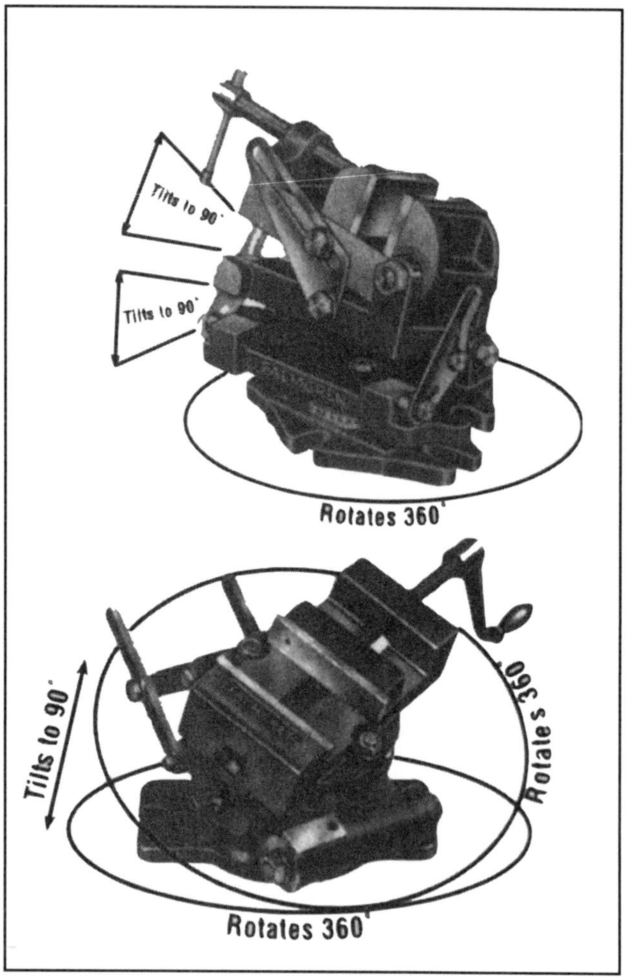

Figure 6.19 *A compound-angle vise.*

Tilting Tables

Tilting tables are used to position larger workpieces (larger than what will fit into a normal vise) to specific angles. They have a series of T-slots, similar to those used in the mill table, and the part is secured to the top of the table. Sizes are somewhat limited, but offer sufficient surface area for most projects machined on a turret mill. Tilting tables adjust from 45° on one side to 45° on the other, making for 90° of total adjustment. These tables can handle heavy-duty applications quite easily. Tilting tables are found in the following table sizes: 5" × 7", 7" × 10", and 10" × 15".

Figure 6.20 *A tilting table.*

Swivel Angle Plates

Like the tilting table, the swivel angle plate is used to position a workpiece to a specific angle. T-slots are used to facilitate the clamping of the workpiece. The swivel table should be used for light- to medium-duty applications. They are found in the following table sizes: 5" × 6", 6" × 8", 8" × 10", and 10" × 12".

Figure 6.21 *A swivel table.*

> *To ensure exact angle location for both angle and swivel tables, you will need to check the angle with a shadow graph.*

Tube/Shaft Vise

The tube, or shaft, vise is used to hold a round workpiece. The vise secures the part between two jaws applying clamping force against a V-block. The V-block has two sides – one for small diameters, the other for larger sizes. The vise can be mounted so that the workpiece is clamped from the top side (horizontal plane), or from the sides (vertical plain).

Figure 6.22 This tube/shaft vise is being used to hold a round workpiece for keyway milling.

Work Stops

Work stops are used to limit and/or position a workpiece. Typically, they are affixed to the vise, or are a stand-alone unit. Stops use a rod to limit the travel of a workpiece as it is clamped in a vise. This makes repeatable positioning of multiple parts very easy – an especially useful feature when machining a large number of pieces. Elimination of finding part zero on each and every piece is very welcome indeed.

Figure 6.23 Work stops allow fast and accurate positioning of parts in the vise.

Cam-Locks

Cam-locks are used to hold awkward shaped parts for machining. They come in a variety of styles and configurations. Typically, the cam-lock clamps the side or edge of the workpiece between three points, minimum. Two of the points must be *fixed* to allow the cam-lock to operate correctly.

Figure 6.24 Cam-locks are very useful when edge-clamping irregular shaped workpieces.

Rotary Tables

A *rotary table* is used to position a workpiece to a predetermined number of degrees relative to a radius. This makes the machining of holes and slots very precise. The workpiece is positioned to the correct location by rotating a feed handle. The peripheral edge of the table is scribed in degrees, and the feed handle is marked in minutes. Feed wheels are turned by hand, or with a motorized feed mechanism.

Figure 6.25 A rotary table.

Rotary tables are found in both horizontal and/or vertical configurations. This extends the operational capability tremendously. To further increase the rotary table's usefulness, a footrest (or "foot stock") can be used in conjunction when the horizontal version of the table is used. The footrest is used to support long workpieces. Some units allow the use of *index plates* that can be used to position parts to within a few seconds of an angle.

Commercially available rotary tables are available in several diameters, including: 6", 8", 9", 10", 12", 14" 15", 18", 20", 25", and 30".

Figure 6.26 The rotary table is used to position parts at specific angles for machining.

Figure 6.27 An indexer is used to position a workpiece for angle machining.

5-C Collets

Initially, *5-C collets* were developed for *speed lathes*, sometimes referred to as *chuckers*. The 5-C collet is somewhat similar to the R-8 collet in function and operation – both hold a workpiece by squeezing its jaws against the outside surface. 5-C collet use on the milling machine requires some type of holding fixture. Spin-indexers, collet indexers, and collet blocks are the most common ways to hold a collet. Each indexer performs essentially the same task – to hold and position a workpiece for machining. Sometimes the part is positioned at an angle, or at a specific number of degrees around the central axis of the workpiece, or a combination of both.

Figure 6.28 A pair of 5-C collets.

The majority of 5-C collets are made from hardened tool steel. They can hold diameters ranging from a few thousandths of an inch (or hundredths of a millimeter) to $1^1/_8$" (28.6 mm). Round, square, and hexagon shaped workpieces can also be held with special 5-C collets. SAE and ISO dimensional collets are available for round workpieces, SAE for square and hex sizes only. The beauty of any collet is its ability to hold stock of varying diametrical sizes — within limits of course. Machine stock as received, from the steel mill may have slight irregularities in diameter when produced. Typically, they will be on the large side, not the small. Therefore, the collet must be able to handle these small discrepancies with ease. The jaws of the collet *spring open* slightly larger than the workpiece diameter. This allows the stock to move in, out, and through the collet. The *over* (plus (positive)) and *under* (negative) *size range* of a collet is typically .010" (0.25 mm).

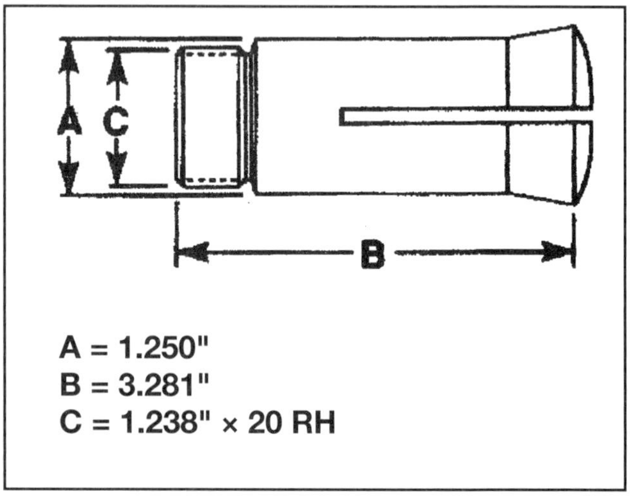

Figure 6.29 5-C collet specifications.

As you will find later in this section, there is a wide variety of collet types to provide work-holding solutions for differing machine operations.

Round Collets Size Chart

Collet Size		Collet Size	
1/16"	.0625"	37/64"	.578125"
5/64"	.078125"	19/32"	.59375"
3/32"	.093753"	9/64"	.609375"
7/64"	.109375"	5/8"	.625"
1/8"	.125"	11/16"	.640625"
9/64"	.140625"	45/64"	.703125"
5/32"	.15625"	23/32"	.71875"
11/64"	.171875"	3/4"	.750"
3/16"	.1875"	49/64"	.765875"
13/64"	.203125"	25/32"	.78175"
7/32"	.21875"	51/64"	.796875"
15/64"	.234375"	13/16"	.8125"
1/4"	.250"	53/64"	.828125"
17/64"	.265625"	27/32"	.84375"
9/32"	.28125"	55/64"	.859125"
19/64"	.29625"	7/8"	.875"
5/16"	.3125"	57/64"	.890625"
21/64"	.328125"	29/32"	.90625"
11/32"	.34375"	59/32"	.921875"
23/64"	.359375"	15/16"	.9375"
3/8"	.375"	61/64"	.953125"
25/64"	.390625"	31/32"	.96875"
13/32"	.40625"	63/64"	.984375"
27/64"	.421875"	1"	1.000"
7/16"	.4375"	1 1/64"	1.015625"
29/64"	.453125"	1 1/32"	1.03125"
15/32"	.46875"	1 3/64"	1.0625"
31/64"	.484375"	1 1/16"	1.0625"
1/2"	.500"	1 5/64"	1.078125"
33/64"	.515625"	1 3/32"	1.09375"
17/32"	.53125"	1 7/64"	1.109375"
35/64"	.456875"	1 1/8"	1.125"
9/16"	.5625"		

Figure 6.30 *Specifications for round workpiece 5-C collets commercially available.*

5-C Square Collet Chart

Collet Size	Decimal Equivalent
1/8"	.125"
3/16"	.1875"
1/4"	.250"
5/16"	.3125"
3/8"	.375"
7/16"	.4375"
1/2"	.500"
9/16"	.5625"
5/8"	.625"
11/16"	.6875"
3/4"	.750"

Figure 6.31 *Specifications for square workpiece 5-C collets commercially available.*

5-C Hex Collet Chart

Collet Size	Decimal Equivalent
1/8"	.125"
3/16"	.1875"
1/4"	.250"
5/16"	.3125"
3/8"	.375"
7/16"	.4375"
1/2"	.500"
9/16"	.5625"
5/8"	.625"
11/16"	.6875"
3/4"	.750"
13/16"	.8125"
7/8"	.8750"

Figure 6.32 *Specifications for hexagon shaped workpiece 5-C collets commercially available.*

Spin-Indexers

The *spin-indexer*, sometimes called a *spindex*, can position a workpiece from 0° to 360° in 1° increments. They use 5-C collets to hold the part.

Figure 6.33 *A spindex. A special wrench (shown) is used to hold the collet during installation and set up.*

5-C Collet Indexers

5-C collet indexers position round, square, or hexagon shaped workpieces in a horizontal position in relationship to the vertical spindle. They use 5-C collets and can accommodate a maximum round stock diameter of $1^1/_8$", and $^3/_4$" for the square and hexagon types.

Figure 6.34 *A 5-C collet indexer fitted with a step collet.*

Indexers can be set up to position the part in an even or odd number of *index* points, or number of degrees. If a workpiece requires machining on three sides, the indexer would be configured to rotate and index every 120°. A simple spin of the handle would align the workpiece from one position to another.

Some indexers are capable of mounting a three-jaw universal or four-jaw independent chuck (with $2^1/_4$"-8 mounting) to hold larger workpieces than the 5-C collets can accommodate. Odd shaped parts can be mounted onto a lathe faceplate. Faceplates can also hold workpiece fixtures for production work, making for quick parts changing.

Figure 6.35 *A 5-C collet indexer set up with a three-jaw chuck.*

5-C Hex and Square Collet Blocks

These two blocks will allow you to machine one, two, three, four, or six locations onto the surface of the workpiece. Blocks are positioned and held with a mill vise. They can be rotated and reclamped to move from one location to another. The hex block is used when a machined surface is required every 60° or 120°. The square block positions the part every 90° or 180°.

Figure 6.36 *The hex and square collet blocks are convenient for performing simple indexing chores.*

The hex and square collet blocks are typically sold as a *pair*. They are an inexpensive solution for the occasional need to machine equidistant-flat surfaces onto the workpiece.

Figure 6.37 *This collet block has a quick-release lever assembly.*

A special nut with a *quick-release* lever assembly connects directly onto the threaded end of the 5-C collet to pull the collet tight. Length of the workpiece is somewhat limited when using the quick-release because the lever closes the end of the collet.

Step Collets

Step collets can hold workpieces up to 4" (101.6 mm) in diameter and are made from steel that is hardened and ground. Three aluminum or soft steel pads, $3^1/_2$" or $4^1/_2$" (88.9 mm or 114.3 mm) diameter, are secured to the collet. The pads are then machined to the desired size and shape necessary to hold the workpiece. Steel pads can be carburized for strength and hardness to increase their life expectancy. Because the pads are removable, the step collet can be re-tooled (new pads installed) for a variety of jobs at a relatively modest cost. The use of a turning machine (engine lathe) may be required to more efficiently machine the desired size(s).

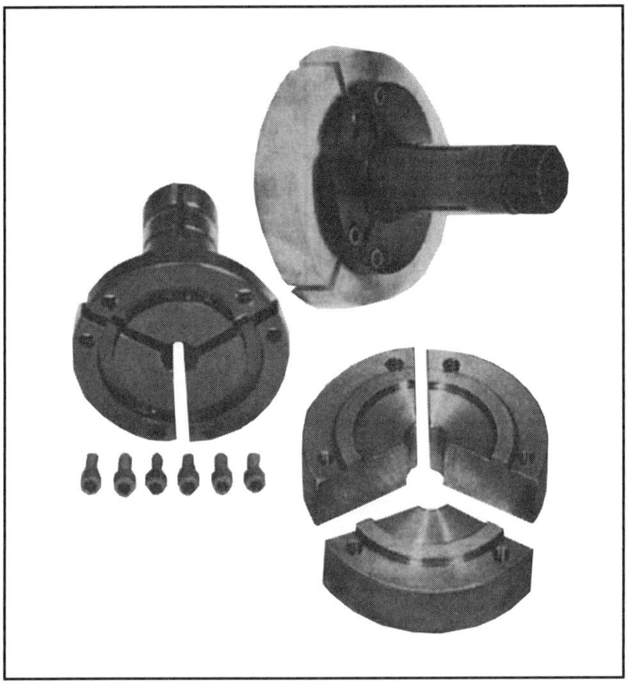

Figure 6.38 This step collet is used to hold a round workpiece. The replaceable soft-jaws can be machined for various workpieces.

Clutch Collets

Clutch collets are similar in use to step collets, they can be machined to hold larger diameter parts for machining. Clutch collets are, however, made entirely from steel. The shank is heat treated for hardness while the head is left soft for machining purposes. These collets are available in 2" (50.8 mm), 3" (76.2 mm), 4" (101.6 mm), 5" (127 mm), and 6" (152.4 mm) diameters. The additional 5" (127 mm) and 6" 152.4 mm) diameter sizes increase the holding capacity over the step collets.

Figure 6.39 A clutch collet.

In order to machine both the step and clutch collets, you will have to mount them into the holding fixture that corresponds with the job you're performing. Three dowel pins must be inserted into pre-drilled holes on the face of the collet before the surface can be machined to size. The recess would be machined oversize if you tightened the collet without inserting the dowels, upon release of the collet, the jaws would *spring back* (open) and the recess would be oversize.

Figure 6.40 A set of dowels are inserted between the jaws to allow machining for the workpiece diameter.

Internal Expanding Collets

As the name implies, the *internal expanding* collet is used to hold and position a workpiece with an interior bore. This can be very useful when holding certain workpieces where a pre-drilled round hole is the only means of mounting.

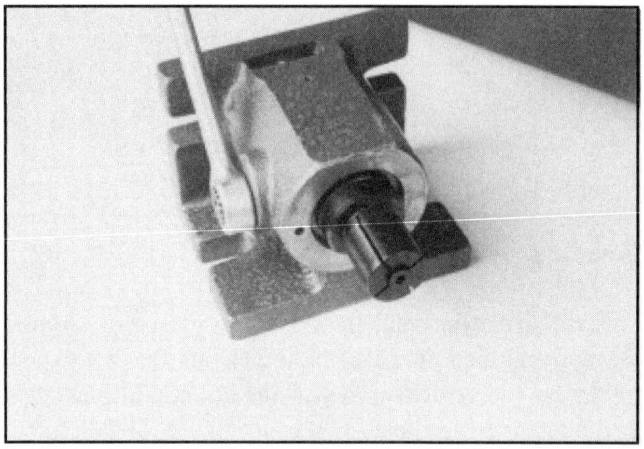

Figure 6.41 *An internal expanding collet.*

Internal collets can be machined to a specific diameter for almost any application. They are available in the following sizes: 5/16" (8 mm) to 9/16" (14.0 mm), 1/2" (12.7 mm) to 1" (25.4 mm), 1/2" (12.7 mm) to 1 1/2" (38.1 mm), 1/2" (12.7 mm) to 2" (50.8 mm), 1/2" 12.7 mm) to 2 1/2" (63.5 mm), 1/2" (12.7 mm) to 2 15/16" (75 mm), 1/2" (12.7 mm) to 3 15/16" (100.0 mm), and 1/2" (12.7 mm) to 4 15/16" (125.3 mm). Collets are available in lengths from: 5/8" (16 mm), 1" (25.4 mm), and 1 3/4" (44.5 mm).

Fixture Mounts

Fixture mounts are used to hold and position a workpiece relative to the centerline of the spindle into which it is mounted. The collet is of a one-piece design. The shank is heat-treated while the head remains soft for machining. The head is carburized to increase hardness.

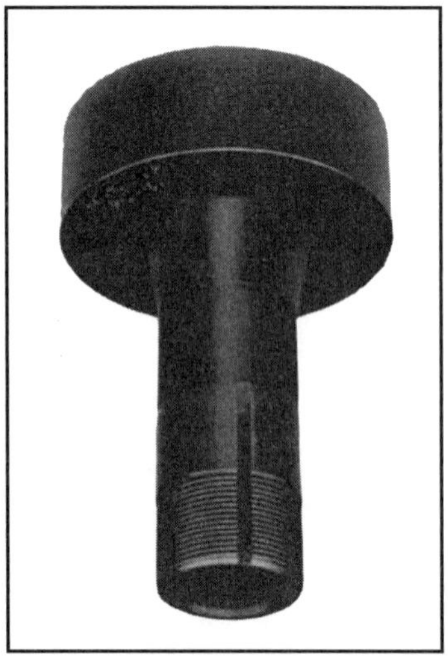

Figure 6.42 *A fixture mount.*

The fixture mount collets have a solid head. It is not used to grip or clamp the workpiece. The head can be drilled and tapped to enable the workpiece to be surface mounted with fasteners. They come in a variety of diameters and head lengths.

Fixture Mount Size Chart

Head Diameter	Head Length
2" (50.8 mm)	1" (25.4 mm)
2" (50.8 mm)	2" (50.8 mm)
2" (50.8 mm)	3" (76.2 mm)
3" (76.2 mm)	1" (25.4 mm)
3" (76.2 mm)	2" (50.8 mm)
3" (76.2 mm)	3" (76.2 mm)
4" (101.6 mm)	1" (25.4 mm)
4" (101.6 mm)	3" (76.2 mm)
5" (127.0 mm)	1" (25.4 mm)
6" (152.4 mm)	1" (25.4 mm)

Figure 6.43 *Specifications for fixture mounts commercially available.*

Index (Dividing) Heads

The use of an *index (dividing) head* is considered to be an *advanced* milling procedure. Index heads are similar in operation to a rotary table, but with a much higher degree of control and accuracy. Instead of a simple degree wheel used to position the workpiece, a handle and set of *index plates* are used to move the workpiece into correct location. The index (dividing) head is held in place on the mill table with a pair of T-slot nuts and bolts.

Figure 6.44 *An index (dividing) head, complete with foot stock and index plate.*

The workpiece is held with a chuck, between centers, or clamped to a flat plate. Chucks are either 3-jaw universal, or 4-jaw independent lathe types. **Note:** When using any chuck to hold parts, be sure to check the piece (after mounting) for concentricity before machining begins.

Placement of the workpiece between centers will require a pair of centers, one in the index (dividing) head and the other located in the foot (tail) stock. Parts held between the centers must be *center drilled* before they can be mounted. This is typically performed with the use of a lathe, as would be the case with a gear blank.

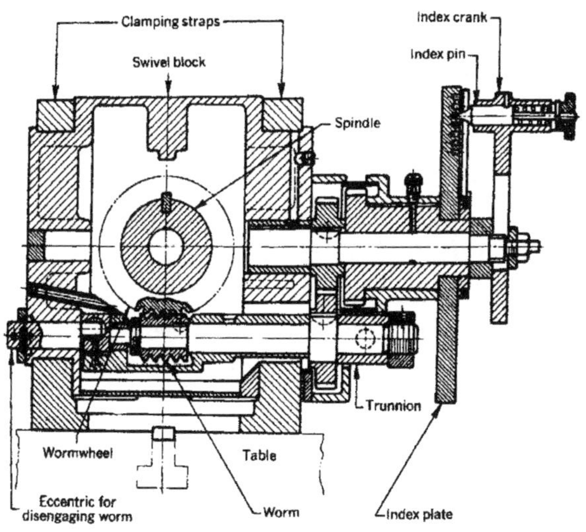

Figure 6.45 *Component parts of the index (dividing) head.*

There are four types of indexing: direct indexing (sometimes called quick indexing), simple indexing, angular indexing, and differential indexing.

Direct Indexing

In this type of indexing a plate with a series of three- hole circles in it with 24, 30, and 36 holes. Any number of holes locations can be divided into any of these three numbers.

Direct Indexing Combination

Divisions Locations	24	30	36
2	X	X	X
3	X	X	X
4	X		X
5		X	
6	X	X	X
8			X
9			X
10		X	
12	X		X
15		X	
18			X
24	X		
30		X	
36			X

Figure 6.46 *This chart is used to determine direct indexing part locations.*

By moving the index (dividing) head to one of the locations listed above in Figure 6.46, you can move the workpiece to the following positions/locations in relationship to a 360° circle: 2 (180°), 3 (120°), 4 (90°), 5 (72°), 6 (60°), 8 (45°), 9 (40°), 10 (36°), 12 (30°), 15 (24°), 18 (20°), 24 (15°), 30 (12°), and 36 (10°). With the number of divisions available, you can move from one position to another very quickly.

Simple Indexing

Simple indexing is, by far, the most commonly used because of the wide range of positions available to you, from 1 to 400. By installing the appropriate index plate you can machine just about any number of locations necessary for the majority of applications. Cutting a gear with 27 teeth is possible, where it is not with direct indexing.

The index (dividing) head uses a worm gear to turn its spindle. It has a ratio of 40:1. Turn the handle 40 turns, and the spindle will rotate one turn. The calculation for simple indexing is as follows:

$$\frac{40}{\text{Number of divisions to be cut}}$$

Equals: Number of turns of the index crank.

Example: $40 \div 2 = 20$ turns of the crank handle

Example: $40 \div 4 = 10$ turns of the crank handle

Example: $40 \div 6 = 6^2/_3$ turns of the crank handle

There are, typically, two or three index plates supplied with the index (dividing) head depending upon the manufacturer. Each plate has a number of holes that are necessary to make the calculations to work out. A pair of *sector arms* are used to position the crank handle at the requisite location.

Plate Style #1

Plate #1	15, 16, 17, 18, 19, 20,
Plate #2	21, 23, 27, 29, 31, 33
Plate #3	37, 39, 41, 43, 47, 49

Figure 6.47 *The Brown & Sharpe index (dividing) head hole patterns plates.*

Plate Style #2

Side #1	24, 25, 28, 30, 34, 37, 38, 39, 41, 42, 43
Side #2	46, 47, 49, 51, 53, 54, 57, 58, 59, 62, 66

Figure 6.48 *The Cincinnati index (dividing) head hole pattern plates.*

Number of Divisions	Hole Circle	Number of Crank Turns	Number of Divisions	Hole Circle	Number of Crank Turns	Number of Divisions	Hole Circle	Number of Crank Turns
2	any	20	**41**	41	40/41	**135**	27	8/27
3	39	13 13/39	**42**	21	20/21	**136**	17	5/17
	27	13 9/27	**43**	43	40/43	**140**	49	14/49
4	any	10	**44**	33	30/33	**144**	18	5/18
5	any	8	**45**	27	24/27	**145**	29	8/29
6	39	6 26/39	**46**	23	20/23	**148**	37	10/37
	27	6 18/27	**47**	47	40/47	**150**	15	4/15
7	49	5 53/49	**48**	18	15/18	**152**	19	5/19
	21	5 15/21	**49**	49	40/49	**155**	31	8/31
8	any	5	**50**	20	16/20	**156**	39	10/39
9	27	4 12/27	**52**	39	30/39	**160**	20	5/20
	18	4 8/18	**54**	27	20/27	**164**	41	10/41
10	any	4	**55**	33	24/33	**165**	33	8/33
11	33	3 21/33	**56**	49	35/49	**168**	21	5/21
12	39	3 13/39	**58**	29	20/29	**170**	17	4/17
	27	3 9/27	**60**	39	26/39	**172**	43	10/43
13	39	3 3/39	**62**	31	20/31	**180**	18	4/18
14	49	2 42/49	**64**	16	10/16	**184**	23	5/23
	21	2 18/21	**65**	39	24/39	**185**	37	8/37
15	39	2 26/39	**66**	33	20/33	**188**	47	10/47
	27	2 18/27	**68**	17	10/17	**190**	19	4/19
16	20	2 10/20	**70**	49	28/49	**195**	39	8/39
17	17	2 6/17	**72**	27	15/27	**196**	49	10/49
18	27	2 6/27	**74**	37	20/37	**200**	20	4/20
	18	2 4/18	**75**	15	8/15	**205**	41	8/41
19	19	2 2/19	**76**	19	10/19	**210**	21	4/21
20	any	2	**78**	39	20/39	**215**	43	8/43
21	21	1 19/21	**80**	20	10/20	**216**	27	5/27
22	33	1 27/33	**82**	41	20/41	**220**	33	6/33
23	23	1 17/23	**84**	21	10/21	**230**	23	4/23
24	39	1 26/39	**85**	17	8/17	**232**	29	5/29
	27	1 17/27	**86**	43	20/43	**235**	47	8/47
25	20	1 12/20	**88**	33	15/33	**240**	18	3/18
26	39	1 21/39	**90**	27	12/27	**245**	49	8/49
27	27	1 13/27	**92**	23	10/23	**248**	31	5/31
28	49	1 21/49	**94**	47	20/47	**260**	39	6/39
	21	1 9/21	**95**	19	8/19	**264**	33	5/33
29	29	1 11/29	**98**	49	20/49	**270**	27	4/27
30	39	1 13/39	**100**	20	8/20	**280**	49	7/49
	27	1 9/27	**104**	39	15/39	**290**	29	4/29
31	31	1 9/31	**105**	21	8/21	**296**	37	5/37
32	20	1 5/20	**108**	27	10/27	**300**	15	2/15
33	33	1 7/33	**110**	33	12/33	**310**	31	4/31
34	17	1 3/17	**115**	23	8/23	**312**	39	5/39
35	49	1 7/49	**116**	29	10/29	**328**	41	5/41
36	27	1 3/27	**120**	39	13/39	**330**	33	4/33
37	37	1 3/37	**124**	31	10/31	**340**	17	2/17
38	19	1 1/19	**128**	16	5/16	**344**	43	5/43
39	39	1 1/39	**130**	39	12/39	**360**	18	2/18
40	any	1	**132**	33	10/33	**400**	40	2/20

Figure 6.49 *This simple indexing chart will provide you with the most commonly used combinations used in the shop.*

Angular Indexing

When specific milling operations require a specific angular relationship, holes, flat areas, etc., you will use *angular indexing* to achieve the desired results. Similar to simple indexing, angular indexing uses the index plates with sector arms to enable you to move the index (dividing) head spindle the necessary number of degrees. Degree measurements of less than 1° are possible with the angular indexing method.

One revolution of the index arm will move the spindle 1/40 of a turn, or 9°. Certainly this is not a fine adjustment when more precise increments are possible with the use of index plates and sector arms.

Differential Indexing

Differential indexing is an indexing procedure that permits the workpiece to move more precisely than when using standard indexing. A set of *differential* gears are added to the index handle to further divide the normal rotation. Think of a 49-hole plate. Now think of moving in a precise increment between the first and second hole of that plate. The differential gear assembly is considered an accessory and must be purchased separately. Each index (dividing) head manufacturer will determine the exact amount of movement that the gear assembly is able to move the spindle. Because of this reason, and the introduction of programmable rotary indexers, differential gear sets are fast becoming outdated.

Programmable Rotary Indexers

Where very high production rates are the norm, the use of a *programmable rotary indexer* is a must. They can be programmed to rotate a 5-C collet to any desirable position at the flick of a switch.

Figure 6.50 *This programmable rotary indexer is capable of setting a workpiece to any number of degrees or location for machining.*

Indexers

Indexers are used to position parts at specific angles. They will typically have a 3-jaw chuck mounted on the rotary part of the unit. Indexers can be mounted horizontally or vertically to position the workpiece. Some indexers have a *quick-release* lever to allow rapid indexing (spinning) from one location to another.

Figure 6.51 *Indexers are used to hold larger workpieces. Some chucks have reversible jaws that can be adjusted to accommodate even larger parts.*

Toggle Clamps

Toggle clamps are used in conjunction with tool plates and machining fixtures. They facilitate the quick installation and clamping of the workpiece. There is a wide variety of clamps available for different clamping requirements. Some are used to pull against the workpiece, while others are used to push against its surface. In either arrangement, the idea is to ensure that the workpiece does not move during the machining process. We will not attempt to describe and illustrate each and every type of clamp, but we will show how toggle clamps are used in several examples. Check with your tool supplier for a list of toggle clamp manufacturers and their specifications for each type.

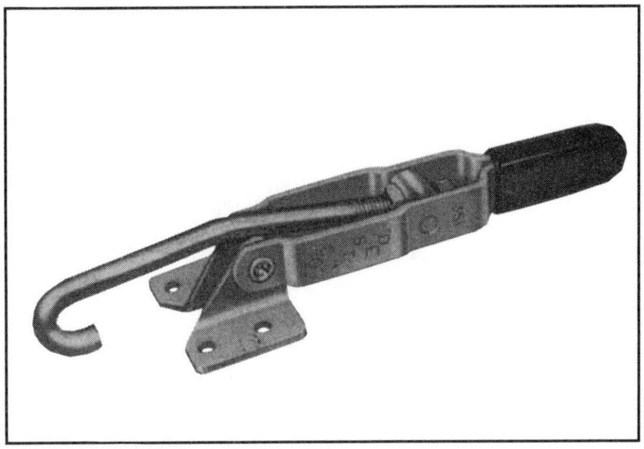

Figure 6.52 *A pull-type toggle clamp.*

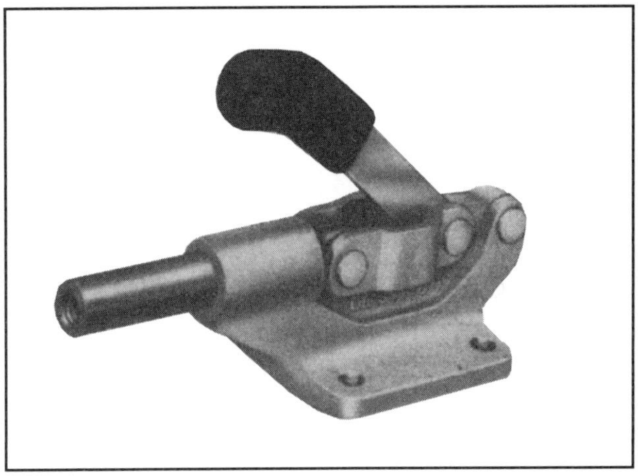

Figure 6.53 *A push-type toggle clamp.*

Figure 6.54 *A bar-clamp toggle clamp.*

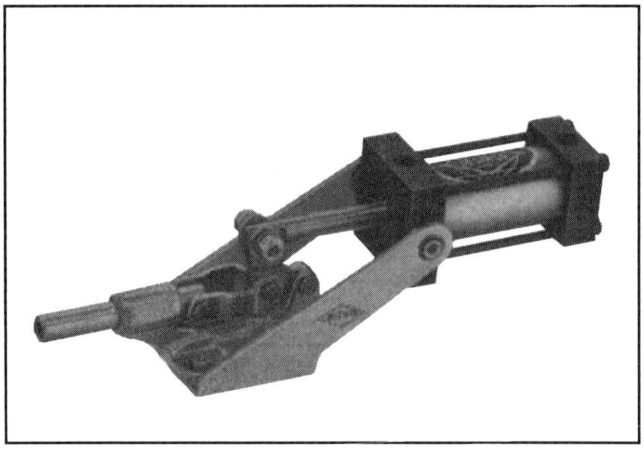

Figure 6.55 *An air-powered toggle clamp.*

Sine Bars

Sine bars are used either for measuring angles, or for locating work to a given angle. They are always used in conjunction with some true surface from which the measurement is taken. As the name implies, the workpiece is positioned to the sine of an angle. The bar uses a pair of what can best be described as *rollers*. The rollers do not rotate, however. They are bolted to the ends of the sine bar and act as rests. The center-to-center distance between the rollers can be whatever you want, however commercially produced sine bars will be $2^{1}/_{2}$", 5", or 10" in length. The rollers allow the ends of the bar to be placed on gage blocks and maintain a consistent distance.

Figure 6.56 *A sine bar is used to position a workpiece in an angular position.*

The sine bar, by itself, is rather useless without a set of **gage blocks.** The blocks are used to raise the end of the sine bar to a predetermined height — the sine of the angle. Gage blocks are sold in sets which contain a number of blocks of different thicknesses. The blocks are typically machined to within plus-minus 50 millionths of an inch.

There are two ways to obtain trigonometric information: with a book or through the use of a scientific function calculator. Let's take a look at how you determine how much gage block you should use for a particular angle. As an illustration, we will use a 5" sine bar and a 30° angle. From the trig tables, or with the calculator, you will find that the sine of 30° is .5000. You will need to multiply the length of the sine bar (5") times the sine (.5000). This equals 2.5000". If you place a .500" block on top of a 2.000" block, then the correct height can be assembled. Next, place one end of the sine on top of the stacked gage blocks. The angle of the sine bar is now 30°. Any part or workpiece placed on the sine bar will also be located at the same 30° position. Simple.

Sine Plates

Like the sine bar, the sine table is used to position the workpiece at an angle. The difference is the size of the table and the ability to bolt or clamp the workpiece directly to the plate. Sine plates are available in two models – single and compound types. The single sine plate places the workpiece in one angle only. The compound sine plate allows the part to be placed in compound (two simultaneous) angles. Commercially available sine plates are available in sizes including: 4" × 4", 6" × 6", 6" × 12", 6" × 18", and 12" × 12". Tapped holes are $^{1}/_{4}$-20 for small size plates, and $^{3}/_{8}$-16 for the larger models.

Figure 6.57 *A sine plate can position a large workpiece at a single or compound angle.* ***Note:*** *This sine plate in this illustration is a magnetic base type and should not be used for machining purposes on a mill.*

Sine Vises

Sine vises are used for direct clamping of the workpiece at an angle. Typical center-to-center distance is 5". These vises are made from tool steel which is hardened and ground parallel and square to .0004". **Note:** Sine vises cannot tolerate heavy machine cuts because of their light-duty side supports.

Figure 6.58 *A sine vise.*

Next . . .

In the next chapter you will learn about coolants and lubricants. Without them, our cutting tool costs would be very high. Coolants remove heat build-up from both the tool and the workpiece. Choosing the correct coolant is another key ingredient — choose the wrong one, and tools will fail prematurely.

Notes:

Chapter 6 ◆ Study Questions

1. Machinist A says that the index (dividing) head requires 20 turns of the handle to make the spindle rotate once, Machinist B says that the handle must be positioned and located to the index plate in order to obtain the correct workpiece location during machining. Who is correct?
   A. Machinist A
   B. Machinist B
   C. Both A and B
   D. Neither A nor B

2. Machinist A says that clutch collets are used to hold extremely small diameter workpieces, Machinist B says that fixture mounts require a set of pins to be inserted between the jaws when machining it to size. Who is correct?
   A. Machinist A
   B. Machinist B
   C. Both A and B
   D. Neither A nor B

3. Machinist A says that spin indexers are accurate to 1°, Machinist B says that they use 5-C collets. Who is correct?
   A. Machinist A
   B. Machinist B
   C. Both A and B
   D. Neither A nor B

4. Machinist A says that the typical bolt size used to clamp parts to the mill table is $^1/_2$", Machinist B says that the table T-slots are $^3/_4$" wide. Who is correct?
   A. Machinist A
   B. Machinist B
   C. Both A and B
   D. Neither A nor B

5. Machinist A says that socket head cap screws are classified as Grade 5 fasteners, Machinist B says that studs used to clamp parts are threaded on each end, or from one end to the other. Who is correct?
   A. Machinist A
   B. Machinist B
   C. Both A and B
   D. Neither A nor B

6. Machinist A says that internal expanding collets grip the workpiece from the outside edges, Machinist B says that step collets hold parts from the inside. Who is correct?
   A. Machinist A
   B. Machinist B
   C. Both A and B
   D. Neither A nor B

7. Machinist A says that some rotary tables may be used in both horizontal and vertical positions, Machinist B says that the rotary table can be positioned in whole degrees and minutes. Who is correct?
   A. Machinist A
   B. Machinist B
   C. Both A and B
   D. Neither A nor B

8. Machinist A says that compound angle vises are used to position the workpiece at two or three angles at the same time, Machinist B says that tube and shaft vises are used, primarily, to hold square workpieces. Who is correct?
   A. Machinist A
   B. Machinist B
   C. Both A and B
   D. Neither A nor B

9. Machinist A says that work stops are used in conjunction with the vise, Machinist B says to use a set of parallels to position the workpiece parallel to the machine table. Who is correct?
   A. Machinist A
   B. Machinist B
   C. Both A and B
   D. Neither A nor B

Chapter 7
Tool Cooling

Tool cooling plays a big role in the overall performance and life of your cutting tools. Not only does the coolant remove heat from the cutter and part, but it also lubricates. The lubrication quality is necessary when cutting certain materials, especially aluminum. Aluminum can form burrs (sometimes called *smearing*) that stick onto the cutting tool. Left alone, the surface finish will become very poor and the size of the cut may increase beyond allowable specifications.

There is a wide variety of coolant types used for cutting tools, however, mineral (petroleum)-base and water-base (synthetic) are the most widely accepted for milling use. Each one has a specific purpose and the selection can make or break a job. For specific applications, contact your local coolant dealer or supplier for a specific recommendation.

Figure 7.1 *Coolants are either mineral (petroleum) or water based (synthetics and water soluble types). Be sure to follow the manufacturer's specifications for application and dilution recommendations.*

There are four general types of coolant application: drip, flood cooling, mist cooling, and high pressure. Drip cooling is one of the most common means to deliver coolant to a workpiece. Brushes used for this job are typically inexpensive and need not be of high quality. It is important to learn how to apply the coolant properly, or you will give the brush a premature and unwanted *hair cut*. Apply the coolant above the flutes of the cutter bit and allow it to drip down and onto the workpiece. Simply placing the brush bristles against the sharp flutes will cause them to be cut off. Also, using a brush with very short bristles in this manner may cause the handle to get stuck between the cutter and the workpiece, jerking it from your hand. This may result in cut fingers or hands if the brush has a metal handle, as is the case of using *acid brushes* for coolant brushes.

Figure 7.2 *Use an acid brush to apply coolant when flood or mist cooling is not necessary.*

Flood cooling uses a low-pressure pump to convey coolant to the workpiece. Excess coolant flows away from the workpiece and into the T-slots. A drain hole is located at each end of the mill table to allow the coolant to return to the storage tank where the pump is located. Coolant may be cooled and cleaned before it is recirculated for use again. Flood cooling is used for low speed spindle applications, high speeds would cause the coolant to be splashed away from the machine and it could not be reused. To say that there would be a mess on the floor would be an understatement. Care must be exercised when handling coolants on and off the machine because most are classified as hazardous waste materials once they have served their usefulness. Upon completion of a job using water-based solutions, be sure to remove all of the coolant from the surfaces of the machine to prevent rust and corrosion.

Figure 7.3 *Flood cooling is used primarily on slow-turning cutting tools.*

Mist coolers are very popular in machine shops for milling machine operations. They do not create as large of a mess as the flood coolant types. The workpiece is partially cooled by an air stream carrying the coolant. This means that much less coolant is used to perform the same cooling function. Spray coolant systems use a small wand-like nozzle to aim and direct the coolant flow. A constant supply of air pressure is also required. Multiple spray nozzles can be mounted where high chip loads are to be expected. Be sure to use some type of vapor vacuum to remove stray mist if you are working in a confined area. Check with the coolant supplier for specific recommendations about their safety precautions.

Figure 7.4 *Mist coolers are used with higher spindle speeds.*

To dramatically increase production and increase tool life, there are high-pressure coolant applicators. These applicators compress the coolant to pressures over 200 PSI (13.6 BAR) and direct the spray at the workpiece. This highly pressurized coolant blows the chips away from the workpiece and the cutter bit and increases surface finish significantly. The down side is the mess. To fully utilize this system the machine must be equipped with side and top shields to trap the overspray. High-pressure coolant applicators work very well on computer numerical controlled (CNC) machining centers with automatic tool changers where operator interface is very limited.

One simple way in which to administer coolant is with a simple spray bottle. The coolant is applied only when and where needed.

Figure 7.5 *A spray bottle is a convenient and inexpensive means to administer coolant.*

For those coolants which must be mixed, mix only that amount that is required for the job. Storage of mixed coolants is a bad idea. Once mixed, coolants start an immediate reaction with air. Left unattended, the mixture will spoil. Spoiled coolants are very easy to identify – they stink very badly. Spoiled coolants should not be reused for cutting, and you should limit your personal contact with them.

Coolants

Now that we know how to deliver the coolant, let's examine the various types and some of their uses. In a perfect world we would not have to use coolants. This would reduce operational costs and eliminate the need to treat them as hazardous wastes (in most cases) when they are disposed of. Yes, coolants do wear out. For a coolant to be effective, it needs to both cool and lubricate the tooling.

Coolants are typically classified as oil-base or water-base. The oil-based types are further classified as mineral or animal based. The mineral types are refined from petroleum-based stock. They receive special fortifiers during the refining process to enhance their interaction with the tooling and workpiece. Using a coolant that causes etching or discoloration to a workpiece is not a desirable characteristic. The majority of mineral-based coolants are applied in pure or 100% form (they are not reduced in strength). These are best applied with a brush or as a flood coolant.

Animal fat-based coolants were among the first coolants used by machinists, because mineral-based types were not readily available. Animal lard is still used for certain materials, especially aluminums. Application is normally with a brush.

Figure 7.6 Animal fat, or lard, is an inexpensive and easy to apply lubricant.

Water-based or synthetic coolants are pretty much the standard in a modern machine shop. They provide excellent cooling and lubrication and can be applied in less than pure form (in most cases). Mixing ratios and strengths must follow manufacturer's specifications for desirable results. Because synthetic coolants are mixed, primarily with water, they need to be stable when they are not used for extended periods of time. The application of the coolant aerates the fluid. Aeration adds oxygen to the solute, which keeps the mixture fresh. Some coolants, if not aerated, will become rancid and spoil. The tell-tale sign, again, is the odor – they stink very badly. When selecting a coolant, be sure to check the bio-stability of the mix and determine how long the mixture can be stored without going bad. The ideal here is to mix minimum quantities, not large amounts. Water-based coolants can be applied by all methods previously discussed. Again, discuss your needs with the supplier for best results.

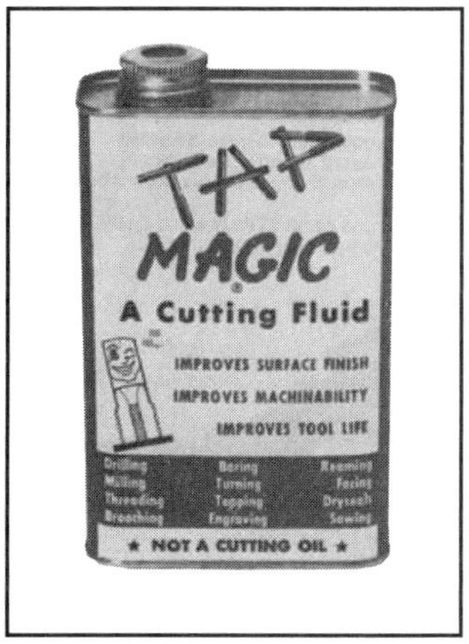

Figure 7.7 Special fluids have been designed and are recommended for tapping of threaded holes.

If you plan to perform tapping operations with the mill, you will want to use a tapping fluid specially formulated for that purpose. They penetrate into the workpiece and tool to make thread cutting much easier than with standard coolants, and they improve the surface finish of the threads. Thread tapping fluids are supplied in an aqueous or paste-like lube, both applied with a brush or straight from the can.

Chemical and Coolant Handling

Always follow manufacturer's handling recommendations for any type coolant. Avoid direct physical contact with your skin and eyes. Use suitable gloves, as necessary, when handling any chemical, and don't forget those safety glasses!

You should always know where *Material Safety Data Sheets (MSDS)* are kept, or ask for them when you purchase coolant. If the unthinkable does happen — take the MSDS with you to the emergency room so that the medical staff will know what they are dealing with. A material safety data sheet file should be set up for all chemicals used in the shop.

Most coolants are classified as hazardous waste after they have served their useful life. To that end, make sure that they are handled and transported within local, state, and/or federal guidelines.

Next . . .

In the next chapter you will learn how to align the mill head to the table and several work holding devices. Alignment tools will include dial and test indicators. Each one is different, yet the same.

Notes:

Chapter 7 ◆ Study Questions

1. Machinist A says that water-based coolants are used straight out of the can, Machinist B says that they must be diluted with water. Who is correct?
   A. Machinist A
   B. Machinist B
   C. Both A and B
   D. Neither A nor B

2. Machinist A says that coolants can be applied with a brush or sprayed with an applicator, Machinist B says that flood cooling should only be used on high speed cutter applications. Who is correct?
   A. Machinist A
   B. Machinist B
   C. Both A and B
   D. Neither A nor B

3. Machinist A says that mineral-based coolants should be mixed with water in a 50/50 ratio for best results, Machinist B says that coolant should be applied above the cutter flutes when applied with a brush. Who is correct?
   A. Machinist A
   B. Machinist B
   C. Both A and B
   D. Neither A nor B

4. Machinist A says that water-based coolants are somewhat corrosive, Machinist B says that they should be completely removed from the machine when you are finished with them. Who is correct?
   A. Machinist A
   B. Machinist B
   C. Both A and B
   D. Neither A nor B

5. Machinist A says that MSDS sheets should be kept on file for all chemicals used in the shop, Machinist B says that MSDS stands for *Material Safety Data Specifications*. Who is correct?
   A. Machinist A
   B. Machinist B
   C. Both A and B
   D. Neither A nor B

6. Machinist A says that oil-based coolants are further declassified as mineral or animal fats, Machinist B says that lard is one type of animal fat used for lubrication and coolant. Who is correct?
   A. Machinist A
   B. Machinist B
   C. Both A and B
   D. Neither A nor B

7. Machinist A says that one of the purposes coolants are used for is to lubricate the cutter bit, Machinist B says that heat will cause a workpiece to increase in size. Who is correct?
   A. Machinist A
   B. Machinist B
   C. Both A and B
   D. Neither A nor B

8. Machinist A says that some coolants need to be aerated from time to time, Machinist B says that this keeps the mixture from becoming rancid or spoiled. Who is correct?
   A. Machinist A
   B. Machinist B
   C. Both A and B
   D. Neither A nor B

9. Machinist A says that used coolants must be treated as hazardous waste once they have served their useful life, Machinist B says that hazardous wastes must be handled with care. Who is correct?
   A. Machinist A
   B. Machinist B
   C. Both A and B
   D. Neither A nor B

10. Machinist A says that coolants never really wear out, Machinist B says that coolants should be replaced periodically. Who is correct?
    A. Machinist A
    B. Machinist B
    C. Both A and B
    D. Neither A nor B

Chapter 8
Machine Head Alignment

Alignment of the mill head to the table is a must if precision work is to be performed. Holes not perpendicular to their surface will cause alignment problems. This is especially true if the hole is to be tapped for threads. Misalignment of the bolt with its threaded recess will force the bolt off-center and create a *hot spot* in the threads that will eventually lead to fastener failure.

When a blueprint specifies a 90° corner, it does not mean 89° or 91°. The head must be aligned to obtain performance values specified in the prints. The mill head must be aligned to the X- and Y-axis of the machine table. This is accomplished with the use of an indicator of the plunger, test, or a specialty type designed for that purpose.

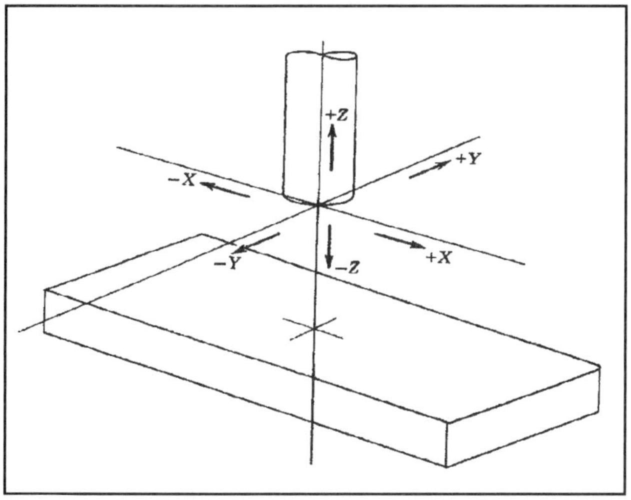

Figure 8.1 *The X-, Y-, and Z-axes of the machine.*

The indicator is secured in, or on, the quill of the machine head depending upon the type of holder used. By sweeping the indicator from side to side (X-axis) the amount of misalignment can be determined. The Y-axis can be checked by sweeping the indicator from front to back on the table. Some machinists use a precision ground round plate 12" to 15" in diameter for this purpose. The plate allows the indicator to be adjusted so that it can be swept (rotated) without interruption of the T-slots in the table. Plus, the larger diameter gives you a wider surface in the Y-axis than what the table will normally allow because most tables are only 9" to 11" wide. The farther out that the indicator point is placed away from the quill center, the more accurate the alignment job will be.

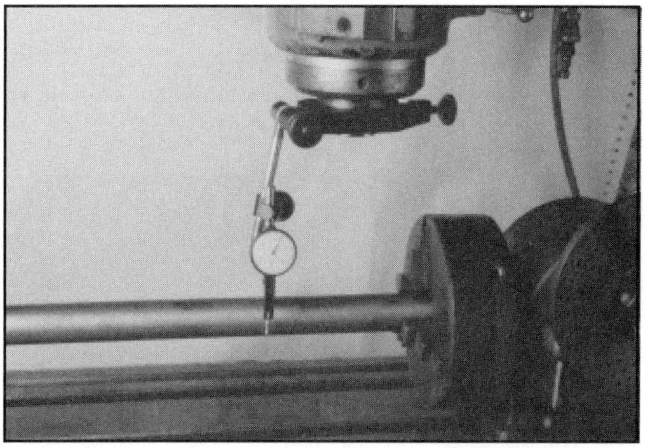

Figure 8.2 *A test indicator fitted to a C-clamp assembly on the nose of the spindle is used to align a dividing head to the machine.*

Figure 8.3 *Here, a plunger indicator is fitted into an R-8 collet to align a shaft vise to the machine.*

The amount of misalignment is determined by the amount of *sweep* that the indicator is put through. The indicator is moved (swept) in a radius around the quill.

The distance that the indicator point is from the quill center will determine the amount of misalignment.

Head Alignment Procedure:

1. Insert or install the indicator holder into/onto the quill. Use appropriate size R-8 collet as required.

2. Install and adjust indicator so that the test end touches the front and back of the mill table (Y-axis).

3. Using the knee table adjustment (Z-axis), feed the table upward until the indicator attains a reading .010". Zero out the indicator.

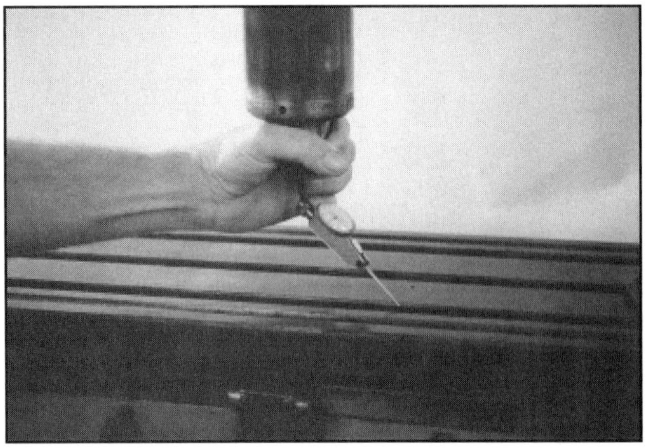

4. Sweep the indicator in a circle across the tabletop. Note indicator readings in both X- and Y-axis.

5. Start your head corrections in the X-axis. Loosen the four clamp bolts on the front of the mill head. The bolts should be snug at this point, not loose.

6. Turn the head alignment adjusting screw on the right side of the mill head with a wrench. Do not make large adjustments.

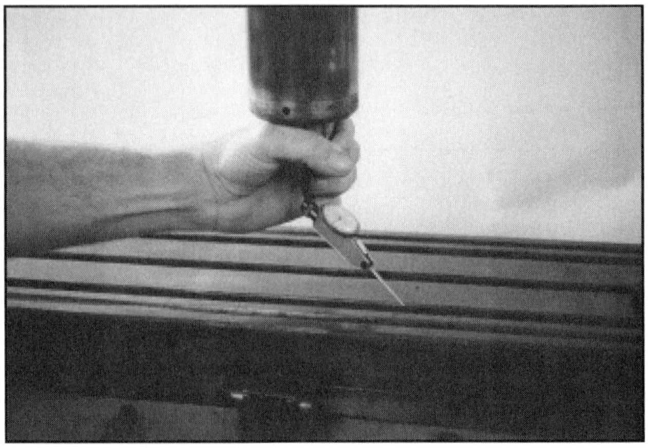

7. Sweep the indicator around again. Note any out-of-level measurements. Make further adjustments as required until the indicator attains a reading of ZERO, or whatever number you have chosen for this purpose.

8. Tighten the four clamp bolts on the front of the machine and recheck your indicator reading.

9. Next, adjust the Y-axis using the same adjustment and checking procedure with the following exception:
   A. Loosen the three clamp bolts located on the side of the mill where the head attaches to the V-ram.
   B. Check and make adjustments as needed.
   C. Tighten the hold down bolts and recheck head for alignment.

Note how the head had to be moved in each axis for the head to become aligned properly. When a high reading is indicated the head must be tilted or raised on that side in order for the indicator to reach the ZERO point. This skill takes time to acquire and will come only with practice.

Vise Alignment

When using a vise to hold the workpiece, be sure to align the vise to the table X-axis. This is especially critical when machining locations on a long narrow part. Should the vise be out of alignment, the machined work will reflect the inaccuracy. Alignment of the vise is not difficult, but you will need to understand what happens when you move it to make alignment corrections. We will examine three separate, but slightly different, vise alignment procedures: a single vise, double (two) vises, and a vise set at a specific angle.

In most machining operations where only one vise is needed to hold a workpiece, the alignment procedure is very straightforward. Simply, clamp the vise to the table and align the fixed jaw to the X-axis. When long workpieces must be held, the use of two vises may be preferable to trying to clamp and align the workpiece to the table. The two vise alignment procedure requires a straight edge long enough to align both vises that will hold the workpiece. And, for parts that must be machined at an angle, the vise can be adjusted with the aid of a machinist's square or precision angle blocks. All of these alignment procedures are part of the machine *set-up* process.

You must learn how to align a vise. It is one of the most basic set-up functions performed be a machinist. The accuracy of the job is tied to the alignment of the vise. If the vise jaws of a 6" (152.4 mm) vise are out of alignment more than .001" (0.025 mm), the part may become out of spec when measured during final inspection. The use of a .0005" (0.00125 mm) reading indicator is highly recommended. Keep the vise aligned to within .0005" (0.00125 mm) from one end of the vise to the other. If higher accuracy is required, install an 18" (457.2 mm) straight edge into the vise and align from end to end to the same .0005" (0.00125 mm) accuracy. This will ensure that the vise is more perfectly aligned than just 6" (152.4 mm) of travel.

Single Vise Alignment Procedure:

1. Carefully place the vise onto the table. Vises are heavy. They must be lifted with care so as not to cause personal injury.

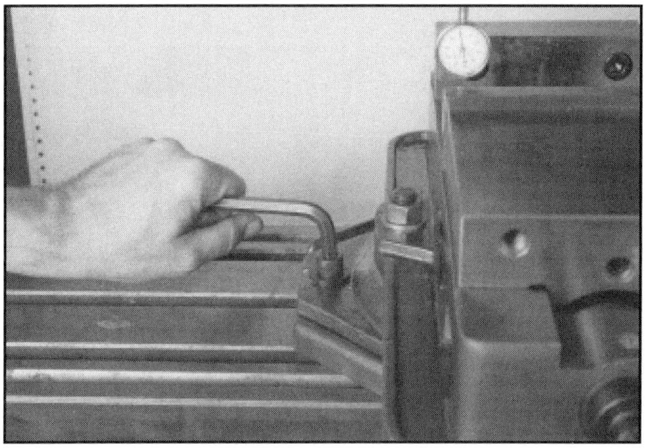

2. Install T-slot nuts and appropriate length fasteners to clamp the vise to the table. Make sure that the fasteners are in the middle of the bolt slots. Always use heavy (thick) washers or flanged nuts. Do not tighten at this point.

3. Install and clamp an 18" straight edge into the vise.

4. Position your body so that you can sight along the top edge of the straight edge and the backside edge of the mill table.

5. This next step will save you several minutes of adjusting time. Move the vise so that the top edge of the straight edge is aligned with the back edge of the table. Tighten the fasteners so that the left side is snug and the right side is somewhat tight. The vise must be able to be moved once the indicator is installed.

6. Install a plunger or test indicator into the quill. Install appropriate R-8 collet as necessary (typically $^1/_4$" diameter).

7. Move the table in the Y-axis until an indicator reading of .020" is attained on the front side of the straight edge.

8. Move the table in the X-axis from one end of the vise to the other (the length of the vise jaws). If further accuracy in the X-axis is required, such as would be the case with very long parts, move the table from one end of the straight edge to the other. Note the difference in the indicator readings.

9. Move the table to the side (X-axis) of the vise where the fastener was tightened snug. Determine if this is the *low* or *high* side of the reading side of the straight edge.

10. Adjust the vise by moving the snug side in or out in the Y-axis to align the vise to the indicator ZERO point. Move the vise to the other side and recheck the reading. Repeat this process until the reading is ZERO. Tighten the vise fasteners.

Double Vise Alignment Procedure:

1. Carefully place both vises onto the table. Vises are heavy. They must be lifted with care so as not to cause personal injury.

2. Install T-slot nuts and appropriate length fasteners to clamp each vise to the table. Make sure that the fasteners are in the middle of the bolt slots. Always use heavy (thick) washers or flanged nuts. Do not tighten at this point.

3. Install and clamp a 36" straight edge, or other straight part, into both vises.

4. Install a plunger or test indicator into the quill. Install appropriate R-8 collet as necessary (typically $^{1}/_{4}$" diameter).

5. Move the table in the Y-axis until an indicator reading of .020" is attained on the front side of the straight edge.

6. Move the table in the X-axis from one vise to the other. Note the difference in the indicator readings.

7. Move the table to the side (X-axis) of the vise where the fastener was tightened snug. Determine if this is the *low* or *high* side of the reading side of the straight edge. Adjust the vise by moving the snug side in or out in the Y-axis to align the vise to the indicator ZERO point. Move the table to the other side and recheck the reading. Repeat this process until the reading is ZERO. Tighten the vise fasteners.

Specific Angle Vise Alignment Procedure:

1. Carefully place the vise onto the table. Vises are heavy. They must be lifted with care so as not to cause personal injury.

2. Place a pre-set (angle) machinists square or precision angle block into the vise. The square or block must stick out of the vise high enough to enable the tip of a test indicator to check the part for alignment, approximately .100" (2.5 mm).

3. Turn the vise so that the angled edge of the square or block is approximately aligned in the X-axis. Some vises may have a base with degrees marked on the peripheral edge.

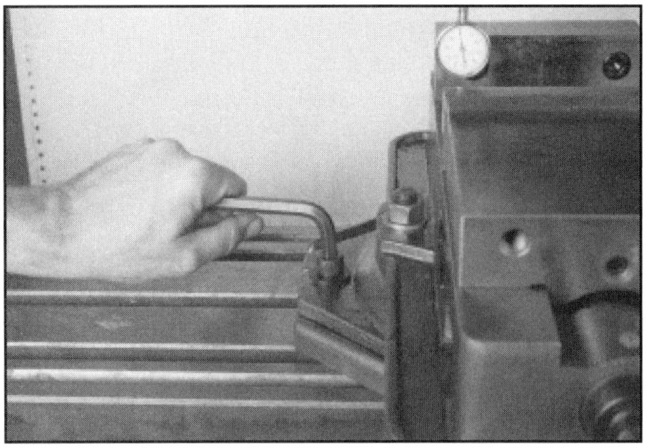

4. Position the vise so that the hold-down slots are aligned with the T-slots in the table. This may mean that the T-slot nuts and fasteners may be placed in different T-slot locations. If you cannot position the vise over a pair of T-slots, use a clamp on the side of the vise to hold it in position. Install T-slot nuts and appropriate length fasteners to clamp the vise to the table.

5. Install a test indicator into the quill. Install appropriate R-8 collet as necessary (typically $^1/_4$" diameter).

6. Move the table in the Y-axis until an indicator reading of .020" is attained on the front side of the square of angle block.

7. Move the table in the X-axis from one end of the square or block to the other. Note the difference in the indicator readings.

8. Move the table to the side (X-axis) of the vise where the fastener was tightened snug. Determine if this is the *low* or *high* side of the reading side of the straight edge. Adjust the vise by moving the snug side in or out in the Y-axis to align the vise to the indicator ZERO point. Move the table to the other side and recheck the reading. Repeat this process until the reading is ZERO. Tighten the fasteners.

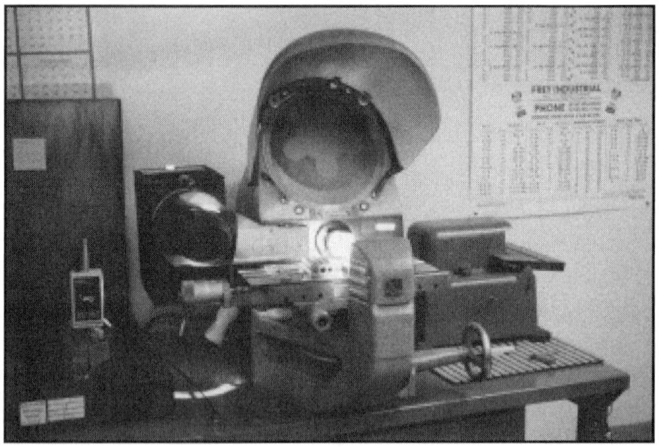

9. In order to confirm that the correct angle has been cut, a test cut must be made onto the workpiece. The workpiece is then removed from the holding fixture and checked with an optical comparator.

Rotary Table Alignment

The rotary table is used to rotate the workpiece to a desired angle. It is most useful when drilling or machining a *bolt circle,* where locations must be a specified number of degrees apart on a common radius. In order to perform this precision work the table must be properly aligned to the mill head. To start, the mill head itself must be properly aligned. If this has not been checked, do so now. There are two widely used methods for aligning a rotary table: either with an indicator or with an alignment plug. The indicator method is the most precise. This method should be studied because it can be used to position a variety of other components and workpieces to the table.

The alignment plug method is used to speed the set-up time. A plug is machined to the hole diameter of the rotary table and installed into the spindle collet. The rotary table is clamped into this position, the table dials are zeroed out, and you are ready to go.

Rotary Table Alignment Procedure:

1. Carefully place the rotary table onto the machine table. Rotary tables are heavy. They must be lifted with care so as not to cause personal injury.

2. Install T-slot nuts and appropriate length fasteners to clamp the rotary table to the machine table. Make sure that the fasteners are in the middle of the bolt slots. Always use heavy (thick) washers or flanged nuts. Do not tighten at this point.

3. Install a tapered alignment tool into the spindle. Some machinists make a special plug that is the same size as the hole in the middle of the rotary table and use it to align the head – it's fast and accurate.

4. Use the quill feed handle and insert the tapered alignment tool into the middle hole of the rotary table. This will reduce set-up time dramatically.

5. Clamp the rotary table to the mill table. Actual alignment is performed with the X- and Y-axis table feed controls.

6. Install a test indicator into the quill with the appropriate R-8 collet as necessary (typically $^{1}/_{4}$" diameter).

7. Move the table up or down (Z-axis) to within a distance of 1"–3" away from the indicator.

8. Move the quill downward and adjust the indicator until the pointer touches the center hole of the rotary table.

9. Rotate the spindle and move the table in X- and Y-axis until a ZERO reading is attained.

10. Lock the table to this position. Adjust the table dials to ZERO.

Note: Get into the habit of always turning the feed handles in the clockwise (CW) direction to eliminate backlash, and set the table feed dials to ZERO once the rotary table is properly aligned. If you used the ZERO to align the head to the rotary table and the power goes out, you will have some real problems getting the head back to the ZERO point.

Use the provided T-slots to locate and secure the workpiece to the rotary table. Use the indicator to align the workpiece to the rotary table if it has a hole in it. You will need to move the part to achieve a ZERO reading. Do not move the rotary table to align the part or you will defeat the whole alignment process of the table.

Index (Dividing) Head Alignment

An index (dividing) head is used to position a workpiece for machining, ranging from horizontal to vertical positions in relationship to the mill table. The majority of index operations are performed in the horizontal position (parallel to the table).

Work is held between centers on an arbor, or with a chuck. In either case, the use of a tail or foot stock would be highly recommended to support the trailing end. Movement or deflection of the workpiece during the machining process will surely render the part out of specification.

Index (Dividing) Head Alignment Procedure:

1. Carefully place the index (dividing) head onto the table. Index heads are heavy. They must be lifted with care so as not to cause personal injury.

2. Install T-slot nuts and appropriate length fasteners to clamp the index (dividing) head to the machine table. Make sure that the fasteners are in the middle of the bolt slots. Always use heavy (thick) washers or flanged nuts. Do not tighten at this point.

3. Clamp the index (dividing) head to the mill table. Actual alignment is performed with the X- and Y-axis table feed controls.

4. Install a plunger or test indicator into the quill. Install appropriate R-8 collet as necessary (typically $^{1}/_{4}$" diameter).

5. Install a round straight shaft between the index centers, or in the 3-jaw chuck.

6. Move the quill downward and adjust the indicator until the pointer touches the edge of the shaft. Move the table (in the Y-axis) until a ZERO reading is attained.

7. Move the machine table from one end of the shaft to the other. Determine the high and low spots.

8. Move the index (dividing) head and tail stock until a ZERO reading is obtained.

9. Tighten the index (dividing) head to the table. Recheck alignment.

5-C Collet Holder Alignment

Pot chucks are used to hold same diameter workpieces typically having a diameter of less than $1^{1}/_{16}$" (27 mm). This chuck uses a 5-C collet, the same as is used on speed lathes. This collet is physically larger in diameter than the R-8 collet, hence the larger diameter capacity. Pot chucks use a handle or pneumatic closer unit to secure the workpiece. 5-C collets have an internal thread to which a *stop rod* can be attached. The stop rod is used to locate the workpiece to a fixed point each and every time the collet is opened and closed. This ensures that each part will be located at the same place for machining.

Figure 8.4 *Alignment of the 5-C collet holder in the vertical plane.*

Figure 8.5 *Alignment of the 5-C collet holder in the horizontal plane.*

Collet holding chucks are available in vertical and/or horizontal designs. Alignment of the horizontal unit is the same as aligning the index (dividing) head. Alignment of the vertical type is like that of the rotary table. Again, use the table feed controls to move the chuck into position.

Spin-Indexer Alignment

The spin indexer is similar in design to the horizontal pot chuck, except that the 5-C collet can be rotated in 1° increments up to 360°. This is a handy feature when milling flat surfaces onto round stock. Start at ZERO and spin the fixture from one location to the next. Alignment of the spin indexer is like that of the index (dividing) head.

Figure 8.6 *Alignment of the spin-indexer in the Y-axis.*

5-C Hex and Square Collet Block Chucks

Block chucks use 5-C collets to hold parts. They come in two configurations – hexagon and square shapes. Parts can quickly be interchanged in them with great ease. Because of their shapes a round workpiece can be machined on six or four sides by simply turning the block from one side to another. They are typically held in a mill vise. Their alignment is dependent upon the alignment of the vise.

Figure 8.7 *Alignment of a 5-C collet block in the horizontal plane.*

Horizontal Milling Attachment Installation and Alignment

The installation and alignment of the horizontal milling machine attachment is critical in order to produce acceptable parts. At first thought, one might imagine that the parts are self-aligning – not so! Each one is interdependent upon the other.

Horizontal Milling Attachment Installation and Alignment Procedure:

1. Place the right angle head into the vise.

2. Position the alignment slot of the right angle head facing towards the front of the machine.

3. Align the alignment pin in the spindle facing forward and lower the quill into the right angle head attachment.

4. Be sure that the alignment pin is positioned correctly, and then tighten the drawbar. **Note:** Some right angle attachments will require the use of a longer drawbar, otherwise use a ratchet and extension to tighten the drawbar.

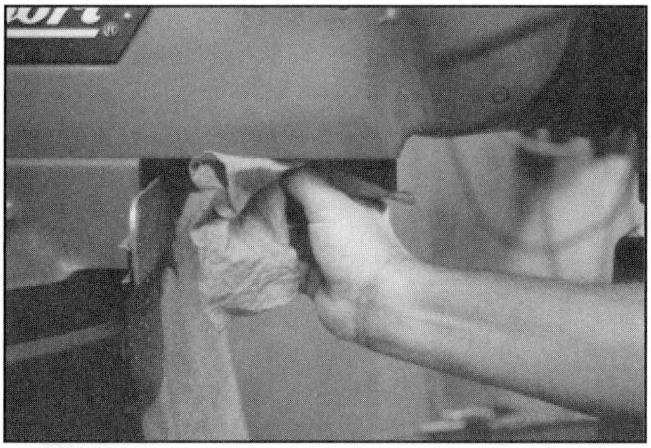

5. Clean the dovetail ways on the over-arm.

6. Install and tighten the pre-assembled arbor into the right angle attachment.

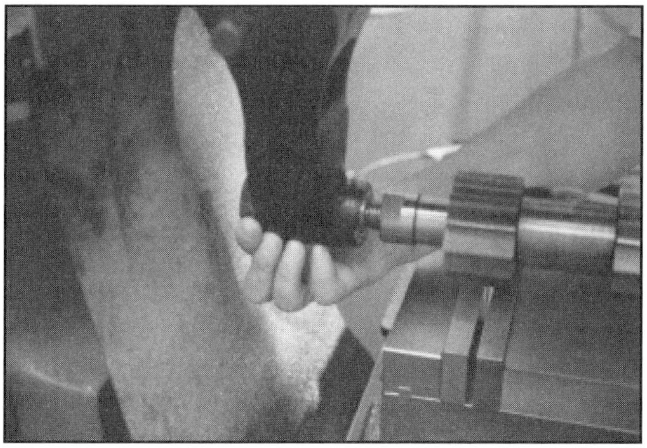

7. Install the arbor support onto the over-arm dovetail. **Note:** This may require movement of the over-arm (front-back) to accommodate the length of the arbor. Move as needed.

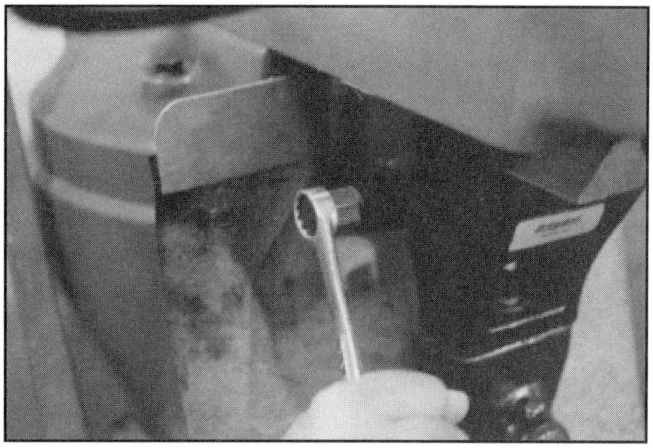

8. Loosen the arbor bearing/bushing located in the arbor support. Slip the arbor into the support. Tighten bearing/bushing until snug – NOT tight.

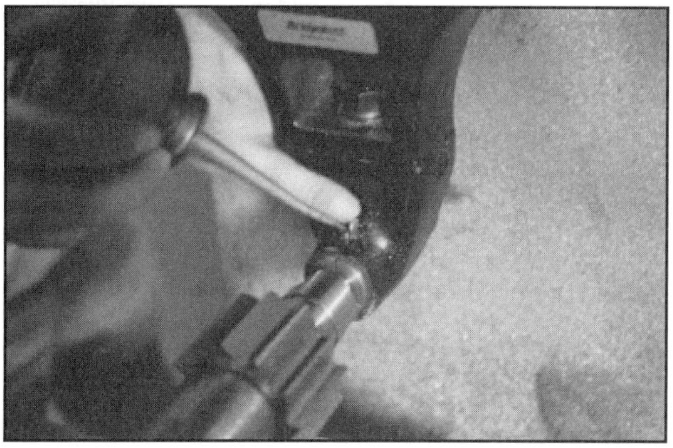

9. Don't forget to lubricate the arbor support bearing/bushing with appropriate lube oil.

10. Place an indicator in the vise and align the right angle head to the machine Y-axis. Tighten rear positioning arm. Re-check alignment after tightening.

11. Now, check the arbor for straightness, adjust as required.

12. Move the indicator and align the arbor in the Z-axis using the knee feed handle. Tighten quill. Re-check alignment.

13. Move the workpiece to the work location. Conventional milling will lessen the strain on the tooling and workpiece.

Work Alignment to the Table

Alignment of the workpiece to the machine table will need to be performed on a one-on-one basis. Not all methods can be covered within the confines of any single book, and the holding methods used by one person may not always be the same as someone else's. But we must all come to some sort of consensus as to where to start. Primary alignment of any of the tools we have already discussed will typically include the use of a precision machinist's square. These squares come in a variety of sizes to fit different part sizes and tooling applications.

By placing the heavy end of the square against the mill table, and the workpiece against the square, you have a pretty good shot at getting the part or tooling quite close. Not many tool makers will produce a tool or fixture that is not square enough to align properly to the table with a machinist's square. Keep this same thought in mind when you make tooling or fixtures – align all surfaces to a perfect 90° corner.

Figure 8.8 *This project is being set up using a machinist's square to align the workpiece to the machine table T-slots.*

Another means to align a workpiece or tooling fixture is to place a rectangular metal bar into one of the T-slots and then locate the piece next to the metal bar. The T-slots are precision machined at the factory using a special one-step machining process.

Some More Thoughts About 5-C Collets . . .

The 5-C collet is available in many sizes including SAE and ISO. They are also available in square and hexagon configurations for holding square and hex stock.

For especially large diameters, *clutch* collets are available. These collets are manufactured in 2", 3", 4", 5" and 6" diameters and can be bored to suit any size within that range. When pulled into position the clutch collet closes around the edge of the workpiece. Irregular shapes can be machined into the face of the collet to suit a variety of odd shaped workpieces.

For those workpieces that cannot be held in position by clamping on the exterior edge, there are *internal expanding* collets. When pulled into position the collet expands against the interior wall of the workpiece. This is an especially nice feature when working with very thin walled workpieces. Internal diameter sizes range from $^{5}/_{16}$" (8 mm) to $4^{15}/_{16}$" (125 mm).

When the need arises for a special shaped 5-C collet, fear

not. Almost any shape can be *burned* into the collet with an *electric discharge machine* (EDM). Once the desired shape is machined onto a carbon or copper electrode, the collet is set up in the EDM machine. By passing regulated current through the electrode and raising and lowering the electrode at calculated rates, the shape is burned into the collet. The result is an exact shape that can then be used to hold the odd shaped workpiece.

Figure 8.9 *An electric discharge machine – EDM (courtesy of AKKT Tool, photo by the author).*

Next . . .

In the next chapter you will learn how to make linear movements in order to position the machine table and quill. First, you must understand Cartesian coordinates and how they are used to help machine a workpiece. Next, you will learn how to find *part zero* with an edge finder or wiggler. When you put them all together, workpieces will be transformed into useful parts.

Notes:

Chapter 8 ◆ Study Questions

1. Machinist A says that vise alignment must be performed with an edge finder, Machinist B says that vise alignment is critical to produce parts that are *on-size*. Who is correct?
   A. Machinist A
   B. Machinist B
   C. Both A and B
   D. Neither A nor B

2. Machinist A says that test indicators can be used to align the mill head to the feed handle, Machinist B says that the mill head must be aligned in the Z- and X-axis in relationship to the machine table. Who is correct?
   A. Machinist A
   B. Machinist B
   C. Both A and B
   D. Neither A nor B

3. Machinist A says that after a vise is aligned to a specific angle, the work must be checked with a profilometer to verify exact geometry, Machinist B says that shadow graphs are able to measure angles cut onto a workpiece. Who is correct?
   A. Machinist A
   B. Machinist B
   C. Both A and B
   D. Neither A nor B

4. Machinist A says that alignment of two vises is not that much different than aligning one, Machinist B says that rotary table alignment can only be performed before the machine head is aligned to the table. Who is correct?
   A. Machinist A
   B. Machinist B
   C. Both A and B
   D. Neither A nor B

5. Machinist A says to use a test indicator to align a rotary table to the machine head, Machinist B says that special sized arbors can also be used to align a rotary table. Who is correct?
   A. Machinist A
   B. Machinist B
   C. Both A and B
   D. Neither A nor B

6. Machinist A says to use heavy (thick) washers to hold the vise or rotary table in place, Machinist B says to use Grade-8 fasteners to hold the vise in place. Who is correct?
   A. Machinist A
   B. Machinist B
   C. Both A and B
   D. Neither A nor B

7. Machinist A says that vise alignment should be checked one more time once the vise has been tightened, Machinist B says that final tightening of the vise may cause the vise to shift slightly. Who is correct?
   A. Machinist A
   B. Machinist B
   C. Both A and B
   D. Neither A nor B

8. Machinist A says that spin-indexers and 5-C collet holders can be aligned by simply inserting an alignment pin into the spindle and the holder, Machinist B says that you should always set the feed handle dials to ZERO after you have aligned the work holder. Who is correct?
   A. Machinist A
   B. Machinist B
   C. Both A and B
   D. Neither A nor B

9. Machinist A says to use a wiggler to align a dividing head to the machine head, Machinist B says that the dividing head can also be aligned with an edge finder. Who is correct?
   A. Machinist A
   B. Machinist B
   C. Both A and B
   D. Neither A nor B

10. Machinist A says that 5-C collets can be machined with an EDM machine if special contours are needed, Machinist B says that horizontal milling machine attachments must be indicated to the work table. Who is correct?
   A. Machinist A
   B. Machinist B
   C. Both A and B
   D. Neither A nor B

Chapter 9
Table and Head Movements

You must understand the Cartesian coordinate (X, Y, and Z) measuring system in order to machine a workpiece properly. Understanding what the coordinates are and how to use them is most important when you move the mill table, knee, or spindle.

Operation of the table is performed with feed handles (one at each end of the table) that move the table from right to left (X-axis). There is another feed handle on the front of the knee that moves the table it in and out (Y-axis). The large handle on the front of the knee controls the up-and-down (Z-axis) movement. For drilling routines (Z-axis), the spindle can be moved in and out of the head with a handle located on the right side of the mill head. Pretty simple, so far.

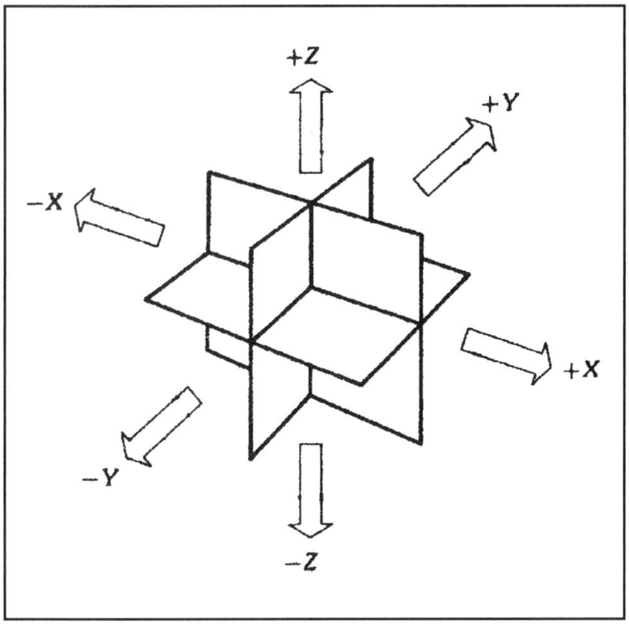

Figure 9.1 *X-, Y-, and Z-axes, the basis of the Cartesian coordinate system.*

The Cartesian coordinate measurement system allows you to create a reference point from which to start machining your part in all three dimensions. Simply stated – length, width, and depth (height). Think of *length* as the side-to-side (right-left) movement of the mill table; the *width* is the in-and-out (front-back) movement; and the *depth (height)* is the up-and-down movement of the knee or quill.

In math, you were first taught how to prepare and read graphs with two dimensions – X and Y. These are called *axes* on the mill. The X-axis is the side-to-side movement of the table; the Y-axis is the in-and-out movement. After you fully understood the X- and Y-axes you were given another axis, the Z-axis. This axis represents height in the Cartesian coordinate system. This corresponds with the up-and-down motion of the mill table or spindle.

You need to establish a *part-zero* once a part has been clamped and secured, either with a vise or a set of clamps. Part zero is the starting point of the Cartesian coordinates X, Y, and Z. It may be necessary to use both *plus* and *minus* Cartesian moves to progress from one location to another.

Part-zero locations will normally be located at the left-front edge of the workpiece. This means that X- and Y-axis movements will be positive, not negative. Z-axis movements will, essentially, always be negative when using the spindle, and positive when using the knee.

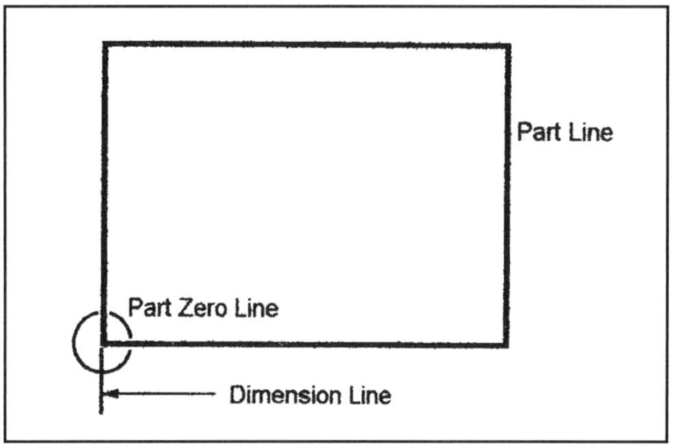

Figure 9.2 *Part-zero will typically have a circle around the corner from which you are to start. Part drawings may not have this symbol. If not, make your own.*

If you are machining a rectangular block, the starting point (part-zero) will be located at a corner. Always return to that point of reference or ZERO point. On a round workpiece, it may be more expedient to reference from the middle, or center. In most cases (with round workpieces) all dimensions will emanate from the central location.

To establish part-zero on a part you will need to use a special tool – a *wiggler* or an *edge finder*. These handy little tools allow you to find the edge of the part and establish part zero. The *wiggler* is, typically, a tool with a ball attached to a small diameter rod held in place with a special chuck and/or collet holding fixture. The *edge-finder* is a mechanical or electric tool used to detect the edge of the part.

The Wiggler

The concept of the wiggler is to rotate a small diameter ball and move it to the edge of the workpiece. Upon start-up of the wiggler the ball will rotate in an eccentric motion. As the tool is slowly moved closer to the edge of the workpiece, the ball becomes more and more concentric to the center of the spindle. At a certain point the ball will appear to have reached concentricity because it will stop skipping in and out against the edge of the workpiece. **Note:** The wiggler must not be rotated at speeds in excess of 500 rpm. Once the ball passes its centerline, centrifugal force will kick the ball off-center. At this point, just before the ball is dislodged from the center, the edge of the part is established.

Figure 9.3 *A wiggler is being used to find part-zero.*

To align the spindle to the exact edge of the workpiece you will need to know the diameter of the ball, which is typically .250". Move the table in or out (depending upon which side of the workpiece the wiggler is positioned) the additional one-half distance (typically .125"). This location will be the part-zero point for that axis. Once the ZERO point is located, adjust the table *feed dials* to ZERO, or press the zero button on the *digital readout* (DRO). More on these shortly.

Wiggler Use Procedure:

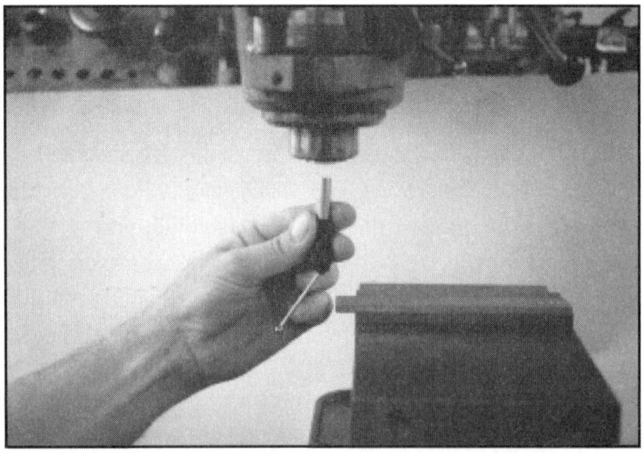

1. Install the wiggler into the mill head with a drill chuck, or with a collet.

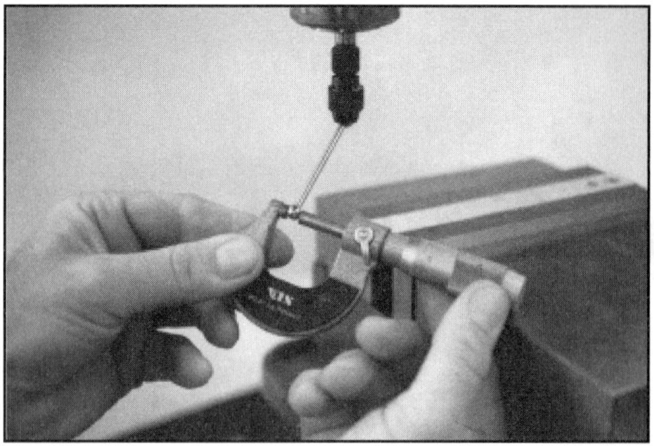

2. Measure the diameter of the wiggler ball. Determine half the ball diameter.

3. Adjust the spindle speed to approximately 500-rpm.

4. Turn on the motor. Rotation is unimportant at this time.

5. The wiggler ball should be running in an eccentric manner. The ball diameter swing should not exceed approximately .500".

6. Slowly move the part towards the rotating wiggler. As you get close to the edge, the wiggler will start to rotate with less and less eccentricity.

7. At the point where the wiggler ball stops running eccentrically, slow the advance of the part towards the wiggler ball.

8. Once the ball just moves away from the edge of the part, stop the advance of the table in that direction. Check and record the location of the machine dial. Repeat the process to ensure that the table will return to the same dial location.

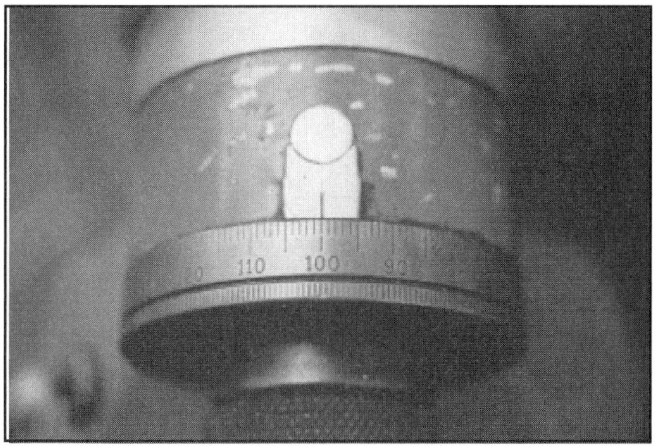

9. Now, advance the table (in the same direction as before) $^1/_2$ of the ball diameter.

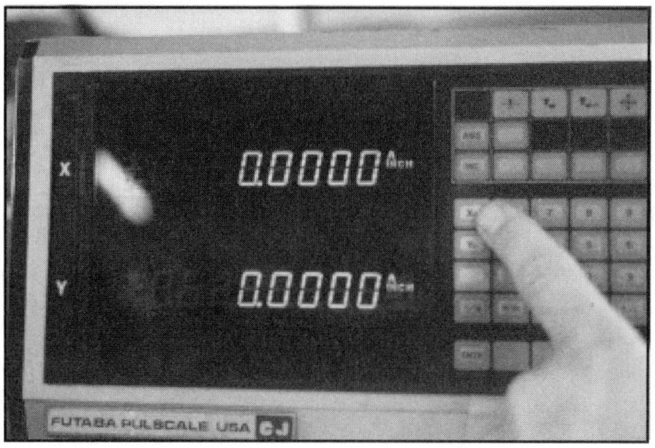

10. Adjust the measurement dials to ZERO, and/or reset the DRO (digital read out) reading.

The Edge Finder

The edge finder is very similar to the wiggler in set-up and operation. Mechanical edge finders must be rotated like the wiggler to determine the edge (part zero) of the workpiece. Instead of a ball, a round pin is used to touch the edge of the workpiece. The round pin will typically have a diameter of .250" or .500". Again, you must calculate half of the diameter in order to complete the final alignment of the table to the ZERO point.

Figure 9.4 *An edge finder being used to find part zero.*

Use of the electric edge finder is easier still. This tool does not need to be rotated – a savings in time. The edge finder uses a ball to detect the edge of the part, typically .400" in diameter. When the ball touches the edge of the part, a light will turn on. This indicates contact and, therefore, the edge location.

The workpiece must be metallic in order for the light to operate. Electric edge finders will <u>not</u> work with composite or non-metallic substances.

Figure 9.5 *Electronic edge finders give you a visual signal when the ball has touched the edge of the workpiece.* ***Note:*** *The workpiece must conduct electricity in order for this tool to work.*

Linear Table Movement

Movement of the table is accomplished with a pair of large feed handles. Turning the dial one way will move the table in one direction, turning it the opposite will move it in the other. The X- and Y-axes feed handles will usually move the table .200" per revolution. However, on some machines the table feed handle will move the table only .100" per revolution. Mills manufactured with metric feed dials will be different. Check your machine to determine the exact specifications. The maximum travel of the table in the X- and Y-axis will depend upon the length of the table and the size of the knee. Consult with the mill manufacturer for exact specifications for this data.

Moving the table is done with the feed handles. The distance covered in one turn of the handle is (typically) .200". When you must move the table more than .200", it will be necessary to add the number of turns required to the total movement amount. This may sound a little confusing at first. Let's look at an example to help clear the air.

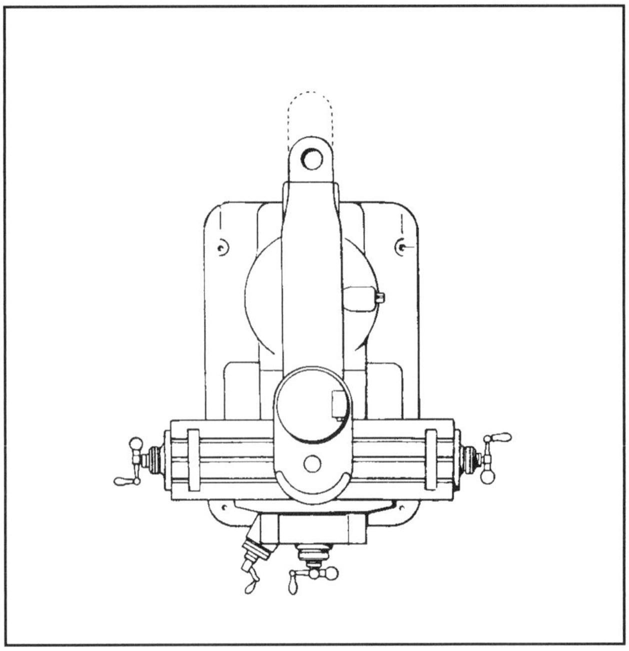

Figure 9.6 *Movement of the table is performed with feed handles, one at each end of the table (X-axis), one for the cross-feed (Y-axis), and one for the knee feed (Z-axis).*

A part must be machined with an end mill for a total distance of 1.250" starting from part zero. If we divide 1.250 by .200 we can see that you will have to turn the handle 6 turns and .050". Six turns will move the table 1.200" (6 × .200 =1.200"). The last .050" is a partial ($^1/_4$) turn of the handle. You must remember to keep count of the turns! Failure to do so, and you may find your part exactly .200" off, or more.

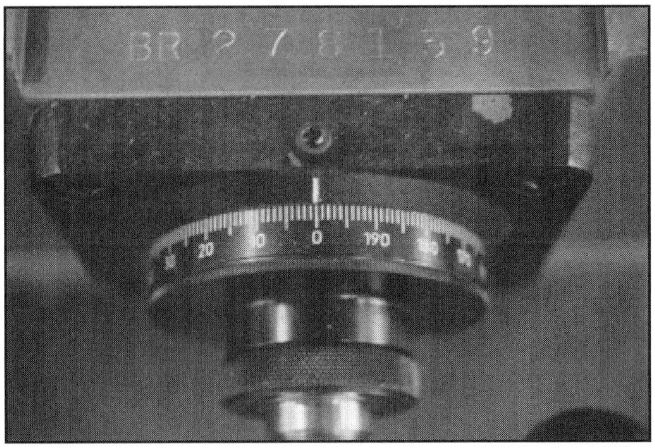

Figure 9.7 *This feed dial moved the table .200" per revolution.*

The knee of the machine (Z-axis) moves .100" per revolution of the feed handle. Maximum rise of the table will normally be limited with a stop-pin or stop-block to ensure that the lead screw does not jam or become dislodged from its nut. Minimum and maximum *rise* (upward movement) and *fall* (downward movement) of the table will depend upon the mill manufacturer. Consult with the mill manufacturer for exact specifications for this data.

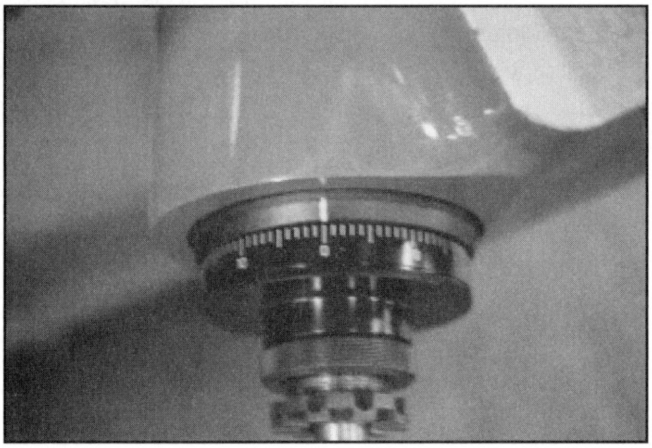

Figure 9.8 *The knee is moved up and down with a large feed handle.*

Movement of the quill is measured with a linear scale attached to the front of the machine head. The maximum travel of the quill will vary from machine to machine, but 5" is the normal maximum travel.

The dials located on the ends (X-axis) and front (Y-axis) of the table are *analog* in nature. The scales have a *minimum resolution*, or measurement, of .001". The mill should be checked regularly for accuracy. The dials can be set to specific locations (measurements), or to ZERO. The dials are held in place with a locking ring or screw.

Backlash

One of the difficulties in making perfect parts on a machine is *backlash*. Standard milling machines will use a *nut and lead-screw* arrangement to move the table back and forth and in and out. There is a certain amount of lash between the two parts when the nut and screw are manufactured. This is necessary to allow for the screw to pass through the nut and not cause it to bind. As the machine ages and wears, the amount of backlash will increase. Adjustments can be made to lessen its effects, but it can never be eliminated.

The majority of work performed on the mill will be at the center of the table. This alone will cause the lead screw to wear more in the center than at the ends. Adjustments to the nut will cause the table to become stiff and require extra effort to move the table from one extreme end to the other.

To reduce backlash you need to remove the standard lead screw and nut arrangement and have *ball-screws* installed. Ball-screws use a very different approach to backlash adjustment; they use a series of balls (under constant pressure) which circulate around the screw. This constant pressure is what eliminates the backlash. Ball-screws are not usually installed as *standard equipment*, but are a very worthwhile option for high-precision machining and reduced machine maintenance.

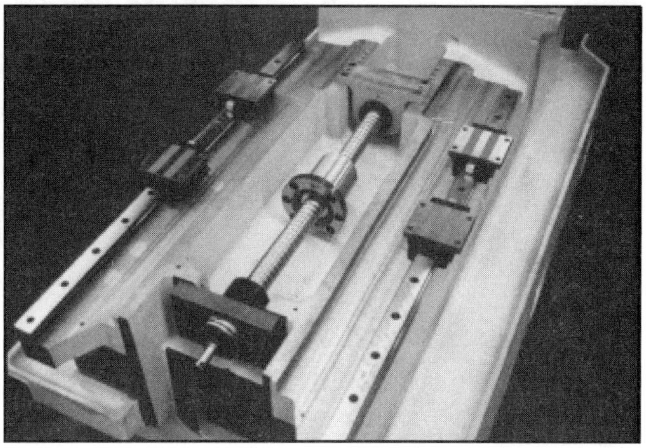

Figure 9.9 *An example of a ball-screw.*

How to Determine Machine Backlash:

1. Turn the table feed handle in a clockwise (CW) direction to remove backlash. Adjust the feed dial to ZERO.

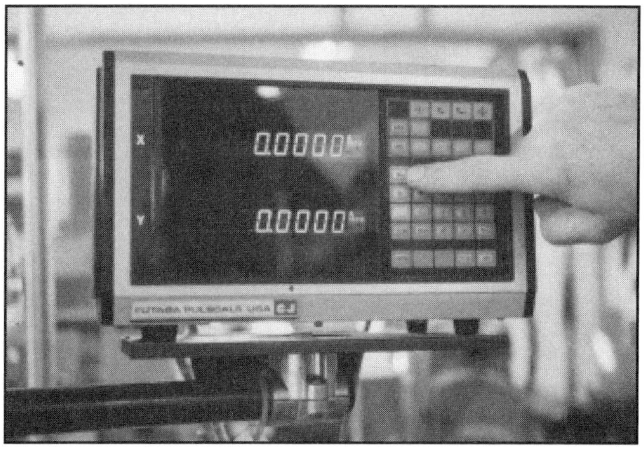

2. ZERO the digital readout.

3. Turn the feed handle clockwise (CW) one full turn (.200").

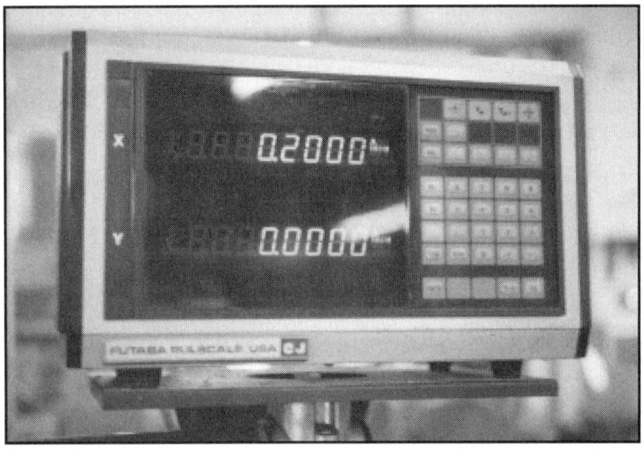

4. The DRO (digital read out) should read exactly .200".

5. Turn the feed handle counter-clockwise (CCW) one full turn (.200") until the dial reads ZERO.

6. The reading on the DRO will indicate the amount of backlash for the table feed. This machine has .0307" backlash in the X-axis.

This same procedure can also be used to determine the amount of backlash on the Y-axis. Knowing the amount of backlash is important in order to produce *on-size* parts.

One of the ways to lessen the harmful effects of backlash is to avoid placing the main load of your work in the center of the table each time you use the mill. Alternate part location by placing them off-center, or in different locations, when you use the machine.

Table Movements Using an Analog (ARO) or Digital Readout (DRO)

In an effort to put an end to the confusion of feed handle revolution counting and backlash concerns, machinists routinely install an *analog* (ARO) or *digital readout* (DRO). Both readouts can measure the X- and Y-axis distance the table is moved very precisely. Readouts can be reset from one measurement or position to another as needed. Readouts for the Z-axis are available as a single add-on unit, or as a member of the X- and Y-axis readout.

Analog readouts use round dials with needle pointers. They engage a roller against the side edge of the table and measure distance as the table moves along that edge from one position to another. The readout must be securely mounted when the table is in motion to ensure exact measurements.

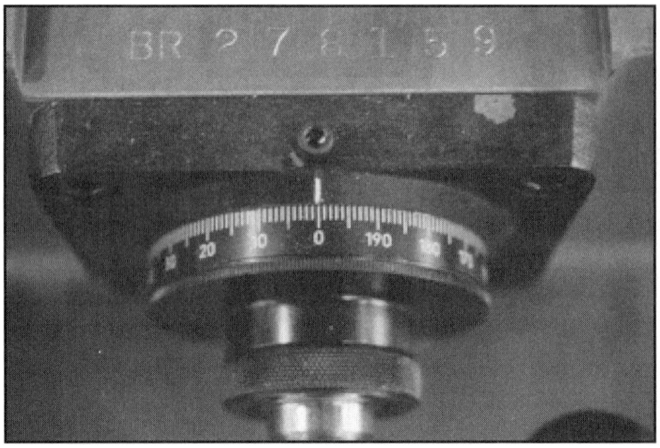

Figure 9.10 *All machines will be fitted with analog readout (ARO) feed dials.*

Digital readout (DRO) displays have just about eliminated the use of analog feed dials and readouts. They display length through the use of *linear encoders*. The linear encoder sends information to a circuit board located within the readout and converts it into information that can be read on the display panel. DROs can be purchased in a number of configurations including instantaneous ISO-SAE conversions, and absolute and incremental measurement systems to name but two. Some DROs have the ability to store limited programming steps and tool types, and can compute bolt circles *on the fly*.

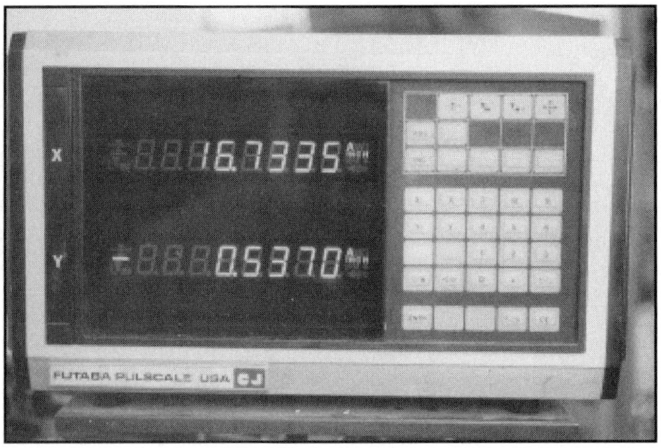

Figure 9.11 *A digital readout registers movement only when the table is actually moved, not when a feed handle is turned. For this reason, more and more machines are supplied with DROs as standard equipment.*

Resolution of the DRO can be found in .0005", .0002", and .0001" (0.001 mm) increments. This is a much closer tolerance than what analog dials (.001") can provide. One of the best reasons for installing a DRO onto a mill is that a DRO does not care what the feed dials say, or how much backlash there is in the machine. Plus, it makes the job go much faster and reduces operator fatigue.

The V-Ram

One of the main reasons why the V-ram mill is so popular is the *V-ram*. This machine component allows the mill head to be moved in and out in relationship to the machine worktable. The movable V-ram allows the placement of large workpieces onto the table for machining.

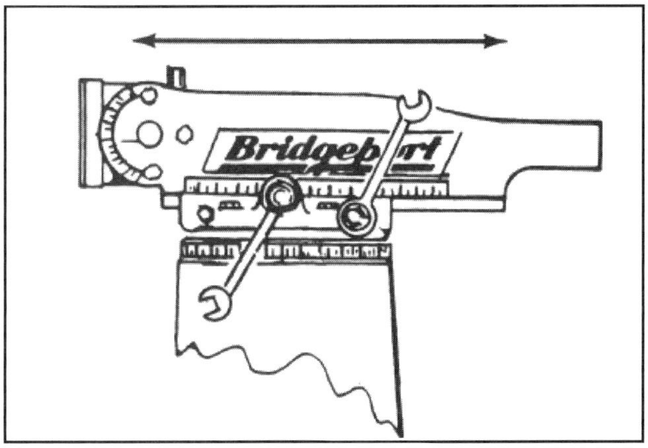

Figure 9.12 *The V-ram is moved with a feed handle mounted on the machine base.*

The V-ram is locked into place with a pair of clamp-bolts. A screw-like jack is used to move the ram in and out. A scale may also be attached to the machine base so that you can return the head to its original position once finished with the oversize project.

Rotating the Head Assembly

The V-ram is located atop a mount that can be rotated in relationship to the machine base. Loosening of the V-ram is accomplished by freeing four (4) nuts – two (2) on either side of the V-ram. Movement of the V-ram can be used to facilitate the machining of oversized parts, or long parts that must hang over the edge of the table. You must use great caution when rotating the head from one position to another. Should you need to tilt the machine head in the future, there may be a possibility of drilling a hole at the wrong angle if the head is not returned to its ZERO point.

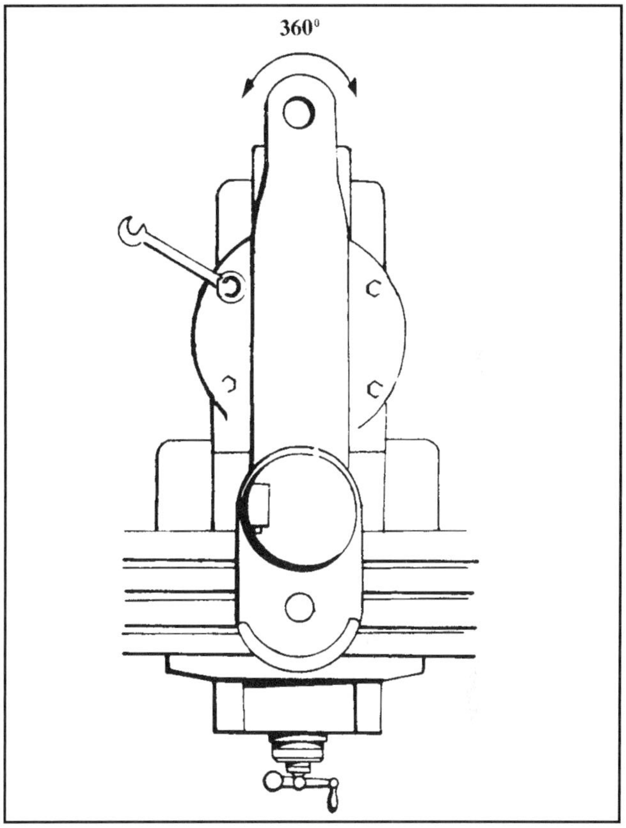

Figure 9.13 *The head can be rotated 360°.*

Tilting the Mill Head

The machine head can be tilted in two directions — side to side and front to back in relationship to the V-ram. There are four locking nuts on the front and three on the right side of the mill head where it attaches to the V-ram. This feature is most handy when drilling multi-angled holes along the length of the workpiece.

The factory-installed protractors used to align the machine head to the table were positioned with as much precision as possible. However, once the head is loosened there is no way to know exactly how the lines were matched during machine assembly. Note the position of the alignment marks for future reference after the head is aligned to the table.

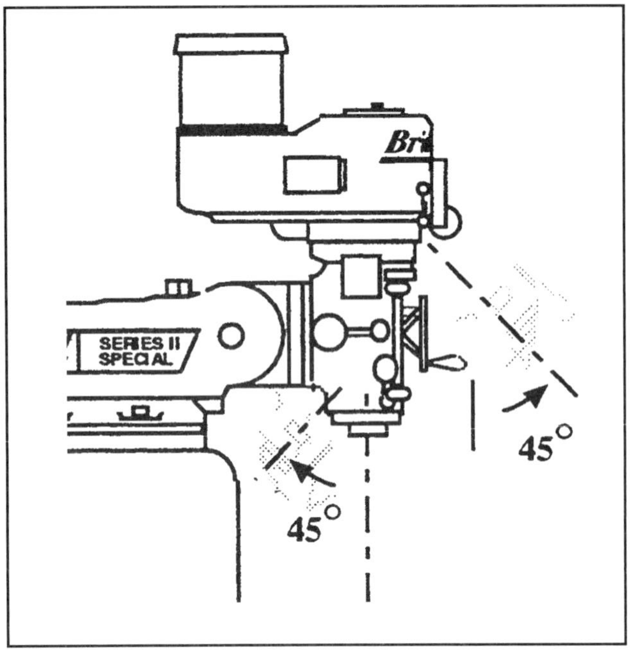

Figure 9.14 *The head can be tilted approximately 90° – 45° forward, and 45° backward.*

Angle alignment of the head must be performed with great precision and may include using a series of angle blocks or a sine bar. Ultimately, angle settings must be verified by placing the machined workpiece upon an optical comparator.

Extreme care must be used when adjusting the mill head. The head is not weight-balanced and is rather top heavy. Tilting the head is typically a two-person job. If you must perform this task alone, place a 3/4" diameter by 12" long steel rod into the machine quill. A vise should also be mounted on the table; it does not have to be aligned — only tight. Bring the vise very close to the rod. As you tilt the head the rod should press against the vise to slow its upward or downward decent. This may require moving the table up or down with the knee feed handle.

Riser Blocks

The use of *riser blocks* can increase the height capacity of your machine. They come in many different heights to accommodate a wide variety of work situations. Sizes include: 5", 10", 15", and 20". Other sizes can be special ordered to suit a particular machining need. Installing a 10" riser will also reduce operator fatigue by lifting the table up where you can see what you are doing. This decreases operator back strain.

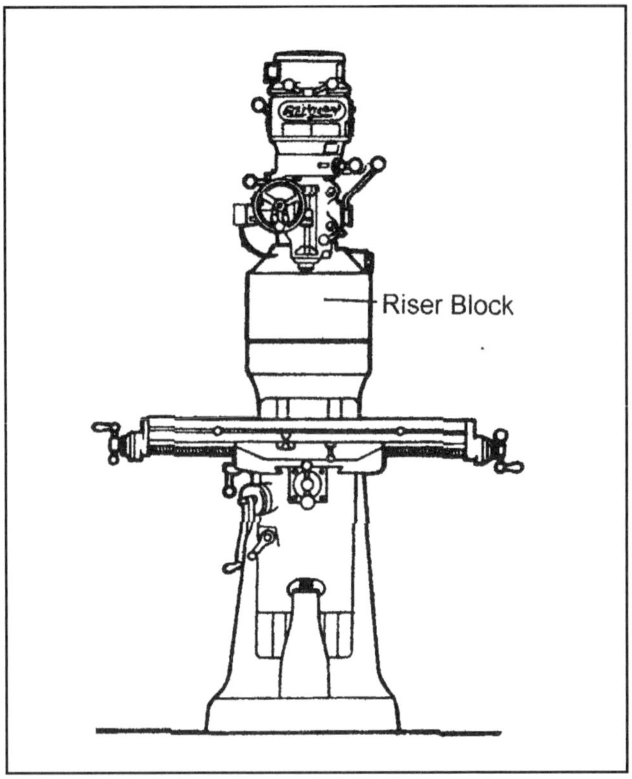

Figure 9.15 *Riser blocks increase the vertical capacity of the milling machine.*

Computer Assisted Controls

The latest generation of turret mills have computerized controls that operate the feed handles. Linear movements are accomplished by teaching the controller where to move, or by programming the moves using *conversational* or *G* and *M* codes. The best part about these machines is that you don't necessarily need to know complicated programming language. However, you had better be very familiar with your Cartesian coordinate measuring system.

The machine itself should have chrome ways and ball-screws in order to stand up to the rigors of motorized table moves. The motors and controllers can be factory installed, or installed by the end user. The choice is yours.

Figure 9.16 *This machine is factory fitted with all the right accessories to make it a real workhorse and increase production 300%-500% over standard machines.*

As stated, the controller can be moved with conversational programming, or G and M codes. The G and M codes are the traditional language used to guide CNC machines. Be aware that not all machines use the same codes. Mills using only G and M control codes will require some tweaking to get them to move when and where you want, but they will do the job very well. G codes are those executable commands that move the tool or part. M codes control motor functions. Enough said for now.

Conversational programming is very user friendly, especially for new users. Unlike having to understand G and M codes, conversational commands are stated, such as "MOVE X 2.500", "DRILL", etc. Remember that conversational programming is unique to the controller and/or machine manufacturer. They are similar to G and M codes, but more specific. For the novice or experienced machinist the conversational programming language will be the first best step to enhancing your skills and knowledge.

The main advantages this type of machine provides are increased production and accuracy. If you have taught (or programmed) the machine correctly, all you have to do is install another workpiece, press RUN, and watch the machine make another part exactly like the one you just removed. All right, maybe that is a little over-simplistic, but it's not far off. Machine manufacturers are making their machines easier to use and operate all the time.

Next . . .

In the next chapter you will learn how to plan a job. As with anything worth doing, you must plan it correctly. Most people fail to plan, not plan to fail.

Notes:

Chapter 9 ◆ Study Questions

1. Machinist A says that a DRO is more precise than standard machine dials, Machinist B says that the table feed dials are calibrated in .001". Who is correct?
   A. Machinist A
   B. Machinist B
   C. Both A and B
   D. Neither A nor B

2. Machinist A says that one full turn of the table feed handle will move the machine table .100", Machinist B says that the knee feed handle moves the table .200" every complete revolution. Who is correct?
   A. Machinist A
   B. Machinist B
   C. Both A and B
   D. Neither A nor B

3. Machinist A says to use a wiggler to determine *part-zero*, Machinist B says that edge finders are used to compute chamfer angles. Who is correct?
   A. Machinist A
   B. Machinist B
   C. Both A and B
   D. Neither A nor B

4. Machinist A says that the Cartesian coordinate measuring system is used to position work in relationship to the X-, Y-, and Z-axes, Machinist B says that the Z-axis describes the linear movement of the quill or knee. Who is correct?
   A. Machinist A
   B. Machinist B
   C. Both A and B
   D. Neither A nor B

5. Machinist A says that backlash in the table must be compensated for, Machinist B says that DROs are typically not affected by lead-screw backlash problems. Who is correct?
   A. Machinist A
   B. Machinist B
   C. Both A and B
   D. Neither A nor B

6. Machinist A says that ball-screws virtually eliminate backlash, Machinist B says that excessive backlash in a machine can be adjusted. Who is correct?
   A. Machinist A
   B. Machinist B
   C. Both A and B
   D. Neither A nor B

7. Machinist A says that riser blocks are used to decrease the maximum distance between the spindle and table, Machinist B says that riser blocks can reduce strain on the operator. Who is correct?
   A. Machinist A
   B. Machinist B
   C. Both A and B
   D. Neither A nor B

8. Machinist A says that CNC assisted machines may be more accurate than manual types, but they produce less work flow, Machinist B says that learning conversational programming language can enhance your skill and knowledge. Who is correct?
   A. Machinist A
   B. Machinist B
   C. Both A and B
   D. Neither A nor B

9. Machinist A says that the mill head cannot be moved or positioned to angles other than ZERO, Machinist B says that the mill head can be positioned *off-axis*. Who is correct?
   A. Machinist A
   B. Machinist B
   C. Both A and B
   D. Neither A nor B

10. Machinist A says that G and M codes are part of the *conversational* programming language, Machinist B says that G and M codes are used to program a CNC machine. Who is correct?
    A. Machinist A
    B. Machinist B
    C. Both A and B
    D. Neither A nor B

Chapter 10
Planning for Projects

?lanning for Projects

The need for planning and preparation for a mill project is ery high. You cannot very easily step up to the mill, turn on the witch, and start machining. So far you have learned a great any procedures used to prepare the mill for work, tool identiication, and how to hold the workpiece – now it's time to learn ow to coordinate a job.

When preparing for a trip, you will make plans — look at naps, make transportation arrangements, book hotel accommoations, plan to see the sights, etc. The same is true for the planing needed to complete a machining project. Only in this case ou will need to look at the drawing or blueprint for specificaons, develop a process sheet, acquire the material for the job, ocate and purchase the tooling, select hold-down tooling, and, inally, perform the work.

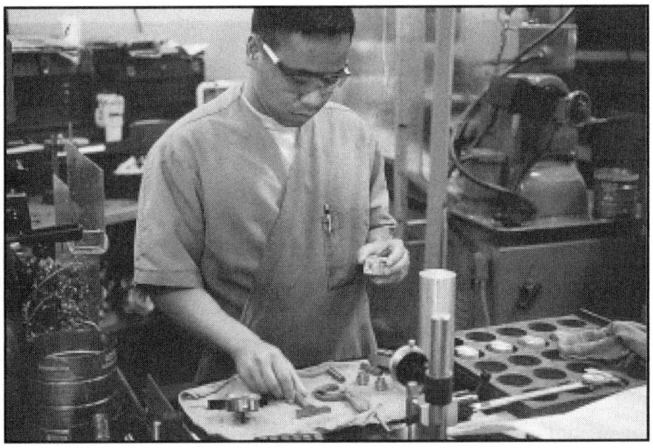

igure 10.1 *This machinist is laying out his tools in an rganized and logical fashion to reduce wasted time and ncrease his production rate.*

Your first task will be to get the part specifications. This is one by getting a set of blueprints or field drawings. Blueprints re formal drawings generated by hand or with a computer sometimes called *computer aided drafting* or *CAD*). They will ontain specifications and information necessary for you to comlete the project. Items like material type, number of pieces, urface finish, heat treatment, and tolerances will augment the art sizes, hole locations, chamfers, etc.

The creation and use of a *process sheet* will speed the compleion time of any job. Learning how to *think a job through*, visulize machine moves, and how to select tooling will enhance your machining skills and proficiency – all necessary to make you a competent machinist. Process sheets allow you to make lists of machines, tools, and procedural steps for the project. Process sheets can be saved for future reference should you have to perform the same job again. Often, projects are very similar, so the process sheet from one job can be used or modified for another.

The process sheet will contain several pieces of information, including the name of the part, material type, size of the stock to be used, and the date upon which the sheet was created, a list of the necessary machines to complete the project, a list of tooling, and a list of the procedural steps with tool identification, speeds and feeds for each step.

Yes, it does take some time to create a process sheet, but not nearly as much time misspent searching for tooling or trying to decide which step should follow or be next. These sheets save time. Saving time will earn you more money.

Make a list of:
- *Cutting Speeds*
  - *Material Selection*
  - *Tool Selection*
  - *Calculating Spindle Speed*
- *Measurements*
  - *SAE*
  - *ISO*
  - *Conversion Factors*
- *Tool Lineup*
- *Part Holding Tool Selection*
- *Sequence/Steps*

Machine Lists

Making a list of machines will help you organize workflow in the shop. In busy shops, certain machines may be allocated for particular jobs or operations. This makes scheduling of that machine a part of your machining sequence priority. The project may require several machines in order to complete the number of operations necessary for job completion.

Your machine list should be laid out using a logical progression of sequences. Don't list the cut-off saw last if it is the first machine used to cut the stock to size.

Learn to develop a code, or shorthand, when listing the machines, e.g., Machine #1 – cut-off saw = M1, #2 – horizontal mill = M2, #3 – surface grinder = M3, etc. You may want to make a list of all of the machines used in the shop and assign a machine number to each of them. This will cut down on confusion and save time. When your shop has more than one machine, such as a lathe, list them with a sub-character — M4a = 9" × 24" lathe, M4b = 14" × 36" lathe, etc. Tooling can be identified with a T code number — T1 = #3 center drill, T2 = #7 drill, T3 = $^{1}/_{4}$-20 plug tap, etc..

Tooling Lists

This list should include all of the tools necessary for the job, including hand tools, holders, cutters, fixtures, and clamping devices. Many machinists will include the hand tools necessary for a particular job or operation. This is because the majority of machinists have their own tools to work with and take them from machine to machine as they move around the shop. In those shops where tooling is kept in a tool crib and must be checked out, a list is definitely in order.

Be specific about the tooling list. After studying the prints, make a list of those tools necessary to cut, drill, ream, and bore the part. Decide which type of holding or fixturing would be best suited for the machining process. If the part is one that is made frequently, there may be a specially designed holding fixture already made – check it out. For those parts requiring standard tooling (vises, T-slot clamps, etc.), decide which ones will make the job easiest to perform. Remember, time is money!

Figure 10.2 *This machinist is checking a list of machine steps required to complete a job.*

Indicate the size and types of cutting tools necessary. Specify tool material type (HSS, carbide, coatings, etc.); number of flutes (2, 3, 4, etc.); standard or center cutting; and lengths. Tool holders must be identified and sequenced. Differentiate which holders are used for quick-and-dirty jobs (those with loose specifications) and those which require a higher degree of precision. You wouldn't want to use a drill chuck to hold tools if a collet will certainly offer more precise location of the cutter.

Process/Step Numbers

The process or step should be numbered starting with the first thing you should do. Typically, this would be to *get stock*. This may seem like a ridiculous place to start, but it isn't. More and more companies are using *statistical-process-control* (SPC) and ISO-9000 standards by which they are able to follow the progression of a workpiece from initial purchase of the stock, through machining, and finally to a completed product. As manufacturing methods become more standardized throughout th world, SPC and ISO will become the norm by which all prod ucts are produced.

SPC is a wonderful *tool* for the machinist if used properl By tracking each part you produce, you will begin to see whe tooling starts to wear out. This will prevent overusing a too which results in a mistake and the scrapping of the part. Re cording part dimensions may seem like a waste or misuse o time, but truly, it's not.

The actual list of steps and the information you include i your process sheet may be driven (again, by SPC or ISO stan dards), or you may be able to include a *bare-bones* list. Alway make a list of the steps as company procedures dictate.

Many of the steps will describe a cutting routine, maybe wi a drill bit, end mill, or large diameter cutter plate. Regardles you will need to compute speeds and feeds for them. Refer the section on *Cutting Tools*. Knowing how to compute the co rect speeds and feeds will save money and aggravation.

Finally, you should include a final inspection sequence o your list. Production machine shops will employ inspectors perform routine and random inspections. Parts may be inspecte in batches or as a complete lot. Typically, you will need to pro duce a *first article* for inspection before you are allowed to jum into the middle of a long production run. The first article woul be the first part you produce once all of the tooling and machir ing operations are refined and set up.

Small shops will generally perform inspections *in-house*. On person may be designated as the *inspector* and it will be up him or her to check parts for size, fit, and finish. Their decisio will be final.

You will, first and foremost, be responsible for checking you own work. Because you are making the parts, initial checkin of the parts will be your responsibility. Be sure that your mea surement tools are *in calibration* by having them checked pric to starting the job. If your micrometer was .001" off, and th specifications gave a tolerance of ±.0003", you would have a lo of explaining to do to your boss why your parts are off size.

Learning how to create a process and set-up sheet will als prepare you for advanced machining projects using *compute numerical control* (CNC) machines. They use a programmin language unique to their application. If you understand th thought process required for the process sheets, learning to un derstand the programming will become a logical extension.

Process and Set-Up Sheet

Page ___ of ___

Part Name: ______________________________ **Date:** ___/___/___

Material Type/Alloy: ____________ **Size:** ____________ **Heat Treatment:** ________ **Surface Finish:** ____ R_a

Machines:

M-1 ______________________ M-4 ______________________

M-2 ______________________ M-5 ______________________

M-3 ______________________ M-6 ______________________

Tooling:

T-1 ______________________ T-6 ______________________

T-2 ______________________ T-7 ______________________

T-3 ______________________ T-8 ______________________

T-4 ______________________ T-9 ______________________

T-5 ______________________ T-10 ______________________

Process/Step	Tool	Speed	Feed
1			
2			
3			
4			
5			
6			
7			
8			
9			
10			
11			
12			
13			
14			
15			
16			
17			
18			
19			
20			
21			
22			
23			
24			
25			

Machinist: ______________________

Notes:

Process and Set-Up Sheet - Continued

Page ___ of ___

Process/Step	Tool	Speed	Feed
26			
27			
28			
29			
30			
31			
32			
33			
34			
35			
36			
37			
38			
39			
40			
41			
42			
43			
44			
45			
46			
47			
48			
49			
50			
51			
52			
53			
54			
55			
56			
57			
58			
59			
60			
61			
62			
63			
64			
65			

Machinist: ______________________________

Sample >> Process and Set-Up Sheet

Page 1 of 1

Part Name: *Bracket* **Date:** 5 / 11 / 98

Material Type/Alloy: 6060T-565 **Size:** 1" × $^1/_2$" **Heat Treatment:** N/A **Surface Finish:** 60 R_a

Machines:

M-1 Horizontal band saw M-1 M-4
M-2 Turret mill M-2 M-5
M-3 Belt sander M-3 M-6

Tooling:

T-1 #3 extra long center-drill T-1 T-6
T-2 #7 jobbers length drill T-2 T-7
T-3 $^1/_4$-20 tap T-3 T-8
T-4 $^3/_8$" jobbers length drill T-4 T-9
T-5 Mill vise 6" jaws T-5 T-10

Process/Step	Tool	Speed	Feed
1 Get stock			
2 Cut stock to length	M-1		
3 Sand edges to deburr	M-3		
4 Center drill holes	M-2, T-1, T-5	2500	.004"
5 Drill tap drill holes	T-2	3980	.004"
6 Tap holes	T-3	Hand	
7 Drill $^3/_8$" hole	T-4	2133	.004"
8 Deburr part	M-3		
9			
10			
11			
12			
13			
14			
15			
16			
17			
18			
19			
20			
21			
22			
23			
24			
25			

Machinist: John Edwards

Notes:

Chapter 10 ◆ Study Questions

1. Machinist A says that you should make a list of the steps necessary to complete a project. Machinist B says that the actual number of steps required to complete a project may be different than the ones that you initially listed on your job or process sheet. Who is correct?
   A. Machinist A
   B. Machinist B
   C. Both A and B
   D. Neither A nor B

2. Why is it important to make a job or process sheet?

3. What types of information should be included in your job or process sheet?

4. How would you distinguish one tool or machine from another with your job or process sheet?

5. Machinist A says drilled hole locations should be center-drilled before they are drilled for more accuracy, Machinist B says that holes should always be through-drilled. Who is correct?
   A. Machinist A
   B. Machinist B
   C. Both A and B
   D. Neither A nor B

6. What does SPC stand for?
   A. Standard Processor Controller
   B. Statistical Process Control
   C. Standard Process Control
   D. None of these

7. Why is SPC used?
   A. To track machinists' mistakes.
   B. To track part sizes and ensure quality control.
   C. To track tool wear.
   D. All of the above.

8. Describe what a *first article* part is and why it is necessary.

9. Understanding how to prepare process sheets can prepare you for CNC programming. True or False?

Chapter 11
Sample Projects

You have learned what the vertical mill is, and about the various tools used to machine a workpiece. Now it's time to put all of the information in this handbook to use. You will start by learning how to clamp and align the workpiece to the machine and then how to machine it.

The method of clamping the workpiece to the table is dependent upon the size and shape of the part. Small pieces can typically be held in a vise or tooling fixture like the *spindex* or a *pot chuck*. Medium to large size parts can be clamped directly to the table. **Note:** If the workpiece must be machined so that the tool cutters must pass completely through the surface (through hole), be sure to elevate or raise the part off of the table to provide sufficient tool clearance.

Anytime that a vise is used to hold a workpiece, be sure to align the vise to the mill table. Failure to do so will cause the part to be machined incorrectly. Re-read the pages containing the vise alignment procedure before you attempt to align the vise to the mill table.

Part Clamping

The use of clamps will be the most practical method to hold a workpiece to the mill table. The shape and size of the workpiece will guide you in your selection of the proper clamp and the fasteners required to secure them. A workpiece with parallel sides will need an appropriate pair of clamps. Small workpieces (less than 2" across) should be held in place with small clamps and step blocks. Larger pieces will necessitate larger clamps and step blocks. Be sure that the clamps are positioned correctly, i.e., parallel to the worktable. Fasteners should be tightened with enough force to secure the workpiece, no more. Keep in mind material type. You cannot clamp a piece of plastic with the same force used for steel or aluminum. The clamp (larger sizes) should grab at least $^1/_4$" to $^3/_8$" of the edge of the workpiece. The clamp bolt (or stud) should be placed as close to the edge of the workpiece as possible without crowding the $^1/_4$" to $^3/_8$" grab area.

Part Alignment

Next comes alignment of the part to the mill table. This can be done with a wiggler or edge finder. If the workpiece is clamped and bolted directly to the mill table, determine which corner you are going to use as part zero and align to it. For workpieces held in a vise, align the vise as previously outlined.

On production jobs a jig or fixture plate will be used to position and hold the workpiece. If this is the case, you will need to align the fixture to the mill table. The workpiece will self-align to the fixture and require no further intervention on your part. Once everything is aligned, it's time to make some serious chips.

Carburetor Jet Holder

Your first project will be a carburetor jet holder. The holder is used to hold and organize fuel-metering jets by size. The block will be made from 6061-T6 aluminum alloy and will have 40 holes drilled and tapped into it. The workpiece should be 1.500" wide, 3.300" long, and .250" thick. Use a vise to clamp the workpiece. Take a look at the drawing at the end of this chapter and pay attention to the hole locations.

Use the following steps to machine the carburetor jet plate:

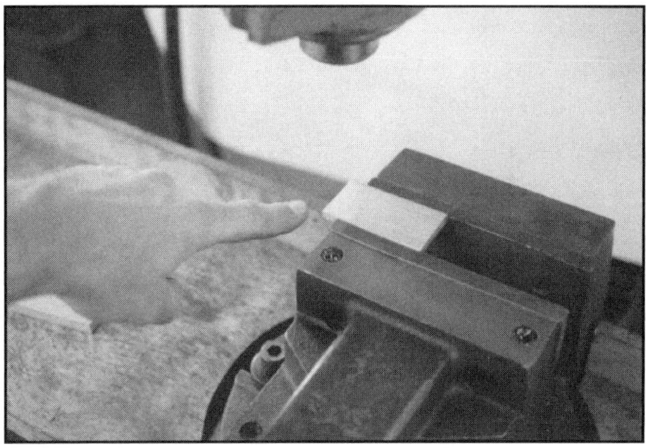

1. Clamp the workpiece in the pre-aligned vise. Ensure that the part is level and parallel to the mill table, use parallels as needed. Also, be sure that you have a corner of the workpiece sticking out of the vise so that you can use the edge finder to find part zero.

2. Find part zero using the lower left-hand corner of the workpiece. Be sure that enough of the edge sticks out of the vise to enable you to find part zero with the wiggler or edge

finder. Zero the machine dials to ZERO and remove any backlash.

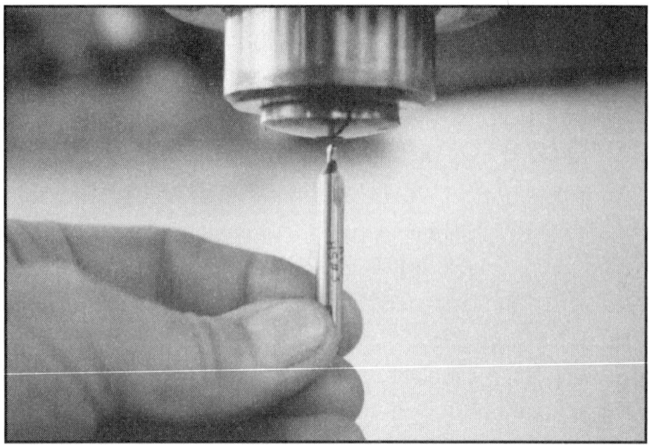

3. Install a 1/4" collet and a #3 center-drill into the spindle.

4. Move the table .300" in the X-axis and -.300" in the Y-axis. Center-drill the first hole location.

5. Move the table another .300" in the X-axis and center-drill that location. Repeat this step again until you have center-drilled 10 more hole locations. Repeat these steps until all 40 holes have been center-drilled.
6. Now, return the table to part zero in both the X- and Y-axes. Eliminate all backlash.

7. Install a #8 drill into collet.

8. Using the same procedures as used in steps 4 and 5, begin drilling.
9. Repeat previous steps until all 40 holes are through-drilled.
10. Remove part from vise.

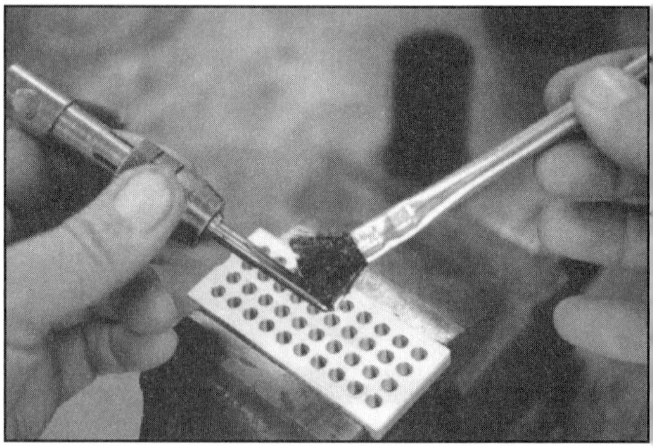

11. Hand-tap holes with 1/4"-28 tap. Be sure to use cutting oil when tapping holes.

2. Tapping can be performed off the machine by placing the workpiece into a bench vise.
3. Deburr part as necessary.

Adjustable Bracket

Brackets are used in many applications and industries for mounting components to one another and to make adjustments. You will make an adjustable bracket with one mounting hole and a slot. You will use a drill bit to drill the mounting holes and an end mill to machine the adjustment slot. You will appreciate the power table feed on this project if your mill is so equipped. See the drawing at the end of this chapter for details.

Use the following steps to complete the bracket:

1. Install and align a vise to the mill table.

2. Lightly clamp the workpiece in the vise.

3. Use a level to correctly orient the workpiece to the machine table.

4. Install a wiggler into the spindle and find part zero on the workpiece.
5. Install a #3 center drill into the machine spindle. Set spindle speed to approximately 1500 rpms.
6. Move the table .500" in the X-axis, and -.500" in the Y-axis.

7. Center-drill the mounting hole. Be sure to use cutting oil on the drill bit.

8. Move the table in the X-axis an additional 2.000" and center-drill.

9. Move the table in the X-axis an additional 3.000" and center-drill.

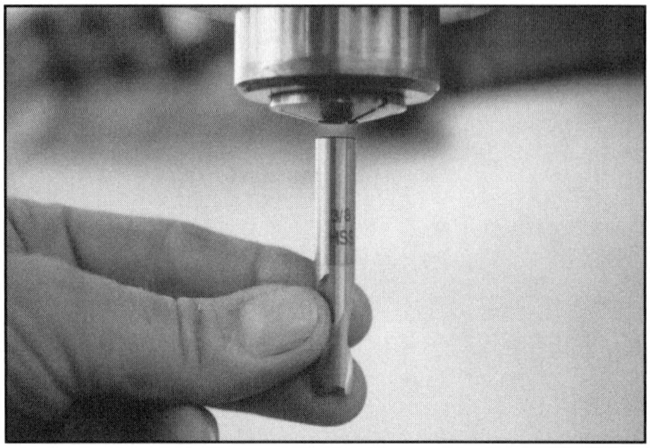

10. Mount a $^{3}/_{8}$" collet and $^{3}/_{8}$" diameter two-flute end mill into the spindle.

11. Prepare to mill the slot into the workpiece 3.000" long. Se the power feed, if installed, to approximately 3" per minute otherwise feed the table by hand. Be sure to use cutting oi on the end mill.

12. Plunge the end mill into the work piece until you have made a through hole.

13. Lock the spindle at this depth.

14. Move the table -3.000" in the X-axis to form the adjustment slot.

15. After milling the slot, raise the spindle and move the table an additional -2.000" in the X-axis.

16. Plunge the end mill into the workpiece until it creates a through hole.

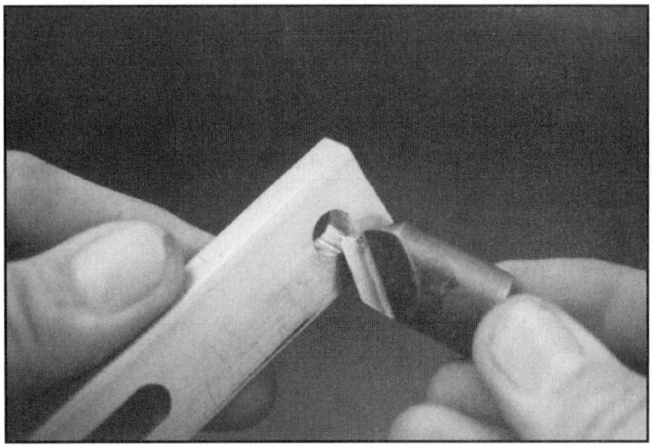

17. Remove the workpiece and deburr all machined edges.

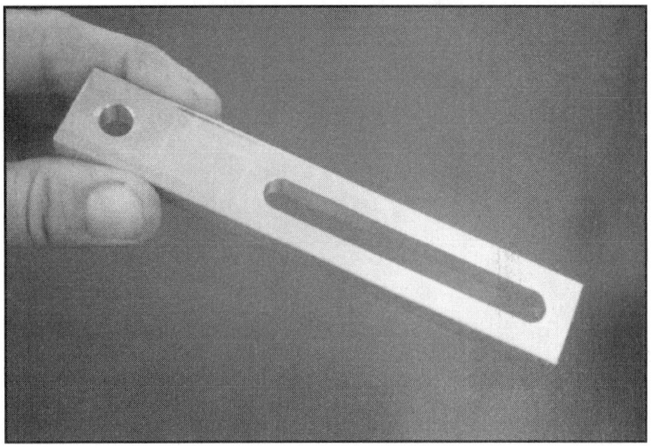

18. This is what the adjustable bracket should look like when you are finished.

Bolt Pattern

First of all, let's look at the difference between a *bolt pattern* and a *bolt circle*. A bolt pattern is a group of holes typically machined into a flat workpiece. The sequence of holes is repeated on other like shaped workpieces. Holes may be placed in symmetrical and non-symmetrical locations, but are always placed in the same locations, e.g., inspection cover plate. A bolt circle is a grouping of holes located in a circular location, each having a common radius, e.g., automobile road wheel.

Our example project will contain six holes located in a rectangular sequence. The holes will be *through* drilled and chamfered to remove burrs and sharp edges. As with all sample projects, make sure that the vise is clamped and indicated to the X-axis before starting. See the drawing at the end of this chapter for details.

Use the following steps to machine the inspection cover plate:

1. Obtain a piece of aluminum stock (any alloy) measuring 2" × 4" × $^1/_4$".
2. Mount the stock into a pre-aligned mill vise with the left side sticking out of the vise approximately $^1/_2$". Check workpiece for parallel alignment (to table) with bubble level.
3. Mount a wiggler or edge finder into the quill and find part zero.
4. Replace the wiggler with a .3125" ($^5/_{16}$") diameter jobbers length drill bit. Install appropriate size collet or drill chuck to accommodate drill bit. Adjust drill speed to approximately 2000 rpm.

5. Move the table .500" in the X-axis and -.500" Y-axis to the first hole location. See drawing.
6. Drill hole.

7. Move the table an additional 1.500" in the X-axis and drill hole.
8. Move the table an additional 1.500" and drill hole.
9. Move the table -1.000" in the X-axis and drill hole.
10. Move the table -1.500" in the X-axis and drill hole.
11. Move the table an additional -1.500" and drill hole.
12. Remove workpiece from vise.

Bolt Circle

Bolt circles are very similar to bolt patterns in the way in which they are laid out, except that the holes are located around a specific radius. This presents the machinist with a problem: how to locate the table in relationship to the cutter bit. The most obvious solution would be to use a rotary table. Simply center the rotary table to the spindle, offset the radius distance, and machine the holes. If you don't have a rotary table, you will need to *trig out* the locations using standard trigonometric functions.

Machining bolt circles is a handy operation to acquire, especially for automotive applications. Relocating bolt holes on axles, hubs, and brake drums is a fairly routine procedure. Like bolt patterns, we machine circular hole location for round inspection plates. See the drawing at the end of this chapter for details.

Use the following steps to machine a bolt circle:

1. Obtain a piece of aluminum, or other suitable stock, measuring: 2" in diameter and $^1/_4$" thick.

2. Clamp and locate the rotary table onto the machine table using a test indicator.

3. Zero the table feed dials.

·. Locate the workpiece and clamp it to the rotary table with suitable clamps and risers. Using risers will position the workpiece so that the drill bit will not drill into the top of the rotary table.

. Now, align the workpiece to the rotary table with the test indicator to within .001" TIR.

). Once the workpiece is centered, move the machine table 15.00 millimeters in the X-axis. This will be sufficient to obtain the specified 30.00 millimeter bolt circle diameter in the specification.
Note: 1.0 millimeter = .03937"

'. Mount a #7 diameter drill bit into the spindle. Set spindle speed to 2500 rpm.

8. Drill the first of six (6) through holes 60° apart. Start at ZERO.

9. Rotate the table an additional 60° (120° total at this point) and drill hole.
10. Rotate table additional 60° (180° total at this point) and drill hole.
11. Rotate table additional 60° (240° total at this point) and drill hole.

12. Rotate table additional 60° (300° total at this point) and drill hole.

13. This is what the hole circle looks like when you are finished.

Keyway Cutting – Parallel Type

Whenever a gear or pulley must be keyed to a shaft, you must machine a keyway into the shaft. The keyway is typically machined with a two- or four-flute end mill. Standard keyway keys are $^{1}/_{8}$", $^{3}/_{16}$", $^{1}/_{4}$", $^{5}/_{16}$", $^{3}/_{8}$", or $^{1}/_{2}$" wide. Other sizes are available; check the actual width of the slot in the gear or pulley before you start to ensure that you are cutting the correct width.

Use the following steps to machine a slot-type keyway:

1. Place the stock into a pre-aligned vise.

2. Check the horizontal alignment of the workpiece with a spirit level.
3. Determine the edge location of the workpiece with an edge finder.
4. Install a $^{1}/_{4}$" diameter end mill into the machine spindle.
5. Move the cutter $^{1}/_{2}$ the workpiece diameter to locate the spindle in the middle of the shaft.

6. Calculate the depth of the keyway and adjust the mill table to that specification.
7. Calculate and set the spindle speed.
8. Advance the cutter down until it just touches the top edge of the workpiece. Lock the spindle into position. Move the workpiece away from the cutter in the X-axis. Adjust the knee feed dial to ZERO. From your calculations, raise the knee to the specified depth.

9. Slowly advance the workpiece towards the rotating cutter.
10. Stop when you have reached the desired length.

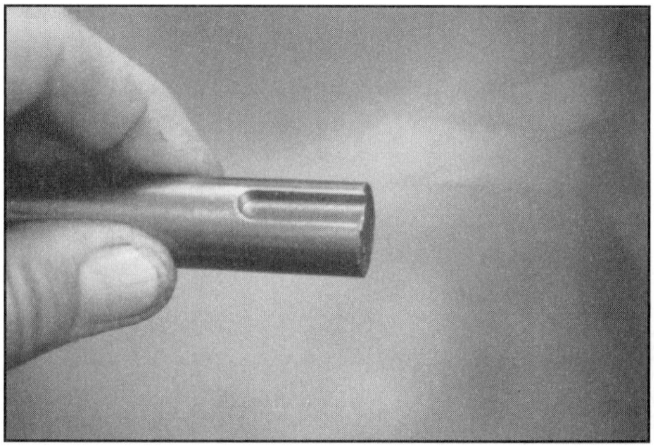

11. This is what a parallel keyway should look like when you are finished. Deburr part with file upon completion of the job.

Keyway Cutting – Woodruff Type

The Woodruff keyway is a half-round slot cut into the shaft and requires a specialized cutter. You must know the diameter and width of the key before you start the machining process.

Use the following steps to machine a Woodruff type keyway:

1. Place the stock into a pre-aligned vise.

2. Check the horizontal alignment of the workpiece with a spirit level.

3. Install the correct size collet and cutter into the machine spindle.

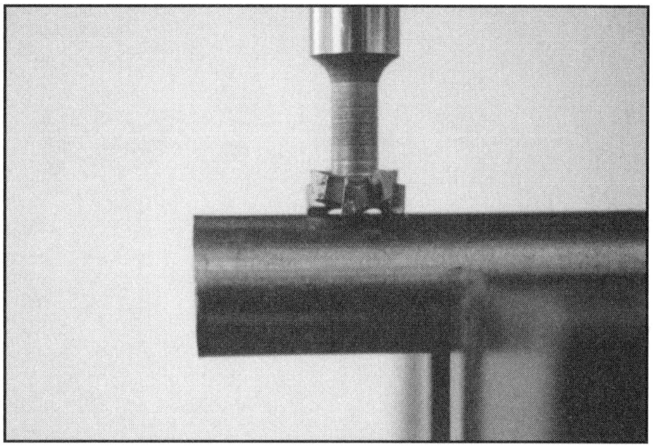

4. Advance the cutter down until it just touches the edge of the workpiece.

5. Move the cutter $^1/_2$ the workpiece diameter ***and*** $^1/_2$ the cutter thickness to locate the spindle to the middle of the shaft.
6. Calculate the depth of the keyway and adjust the mill table to that specification.
7. Calculate and set the spindle speed.

8. Lubricate the cutter. Slowly, advance the workpiece towards the rotating cutter. Stop when the cutter just touches the edge of the workpiece. Zero the machine feed dial.

9. Next, advance the cutter to within .010" of the desired keyway depth. Install the keyway and check protrusion.
10. Cut the final amount required to position the Woodruff keyway at the desired depth into the workpiece.

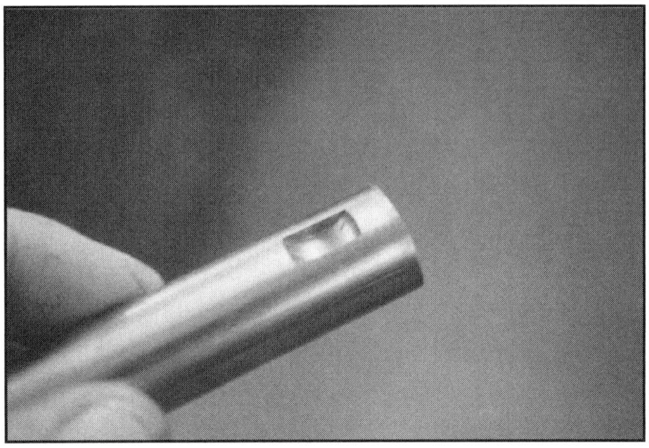

11. This is what the keyway should look like when you are finished.

Timing Belt Gear

This is a beginning gear cutting great project made with an index (dividing) head. The gear blank is made from plastic and can be cut with a simple *form tool*. The tool in the illustrations was made from a worn-out $^1/_4$" two-flute end mill.

In order to complete this project you will need to understand some of the math which relates to this type of gear cutting. The calculations necessary for the gear will be broken down into steps containing simple math.

Timing gears are used in a wide variety of applications, including: automotive, industrial power drives, electronic equipment, washing machines, etc. The timing belt is made of reinforced rubber and comes in several widths, lengths, and tooth spacing. Our project has a $^3/_8$" (L-series timing belt) pitch. This means that the distance from one tooth to another is $^3/_8$", or .375".

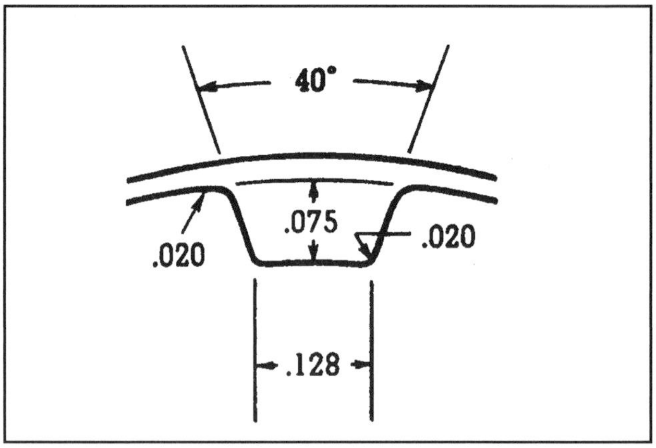

Figure 11.1 *This diagram illustrates the specifications and measurements of an L-series timing belt with a $^3/_8$" pitch.*

The gear is the reverse of the belt tooth configuration. If yo turn a timing belt wrong-side out you will see how this work We must know how big the gear *blank* must be before we ca start the actual cutting process. Our project will have 22 teeth

To start, let's figure out how long a belt with 22 teeth woul be. By multiplying the belt pitch (p) times the number of teet (t) we come up with the length.

p = .375   p × t = 8.250"
t = 22

.375" × 22 = 8.250"

If we connect the ends of the 22-tooth belt together, we forr a circle. But, we still don't know what diameter the gear blan needs to be. However, we do know the length (circumference) i 8.250". If you divide the length (l) times π (3.1416) we will fin the diameter.

l = 8.250"   l × π = 2.626"
π = 3.1416
:
8.250 - 3.1416 = 2.626"

We are getting close, but you need one more piece of info mation, the belt tooth thickness, .045". If you simply subtra .045" from the diameter, you would be wrong. The belt thick ness must be calculated for both sides of the gear, not one. Th brings our total belt thickness to .090" (b).

b = .090"   2.626" - b = 2.536"

2.626" - .090" = 2.536"

As calculated, our gear blank diameter is 2.536". This spec fication, along with the blank width and shaft bore size, can b sent to the lathe department where the blank will be turned to size. Now that we understand the math portion of the project, let's turn our attention to the gear tooth cutting part of this project.

Use the following steps to complete the timing gear:

1. Place the index (dividing) head and foot stock onto the mill table and secure both with appropriate sized T-slot nuts and bolts. Note the finished gear in this illustration. Ensure that the arbor is properly aligned to the dividing head by using an indicator to align the head and foot stocks to the mill table.

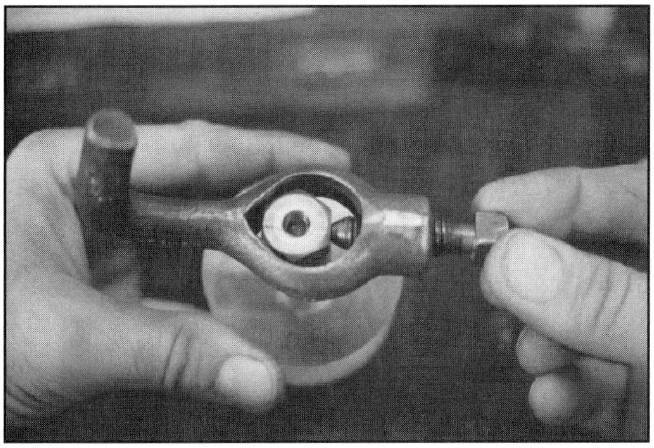

2. Place the plastic gear blank onto the arbor. The arbor used for this project is a simple nut and bolt arrangement that has been center-drilled on each end. **Note:** This may require the use of an arbor press to seat the arbor into the blank. Be very careful when seating the arbor in the plastic gear blank. Excesive pressure will cause the blank to split. Finally, attach a lathe drive dog onto the arbor.

3. Place the arbor between the centers.

4. Tighten the clamp screw against the drive dog. Do not use excessive force on the screw as damage to the drive may result.
5. After installing an edge finder into the spindle, determine where the edge of the gear blank is located in relationship to the center of the spindle. Remove edge finder and move the table to the exact middle of the gear blank, one-half the blank diameter. Lock the Y-axis table lock.
6. Install the form tool and collet into the spindle. Tighten the collet.

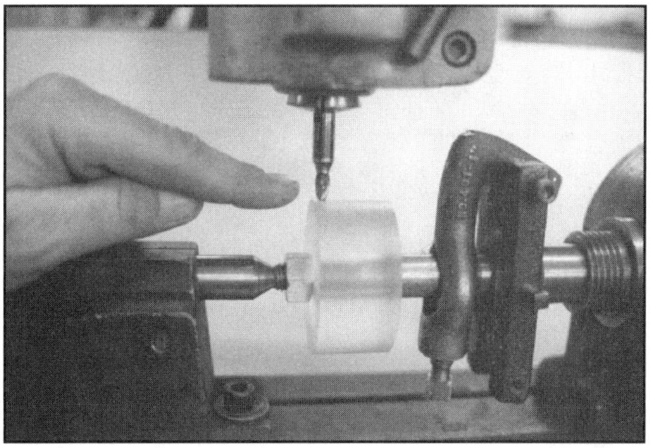

7. Advance the tool down (Z-axis) until it just touches the blank. Lock the spindle in this location.

8. ZERO out the knee scale. All of your Z-axis moves will be performed by moving the knee.
9. Set the spindle speed to approximately 2500 rpm.

10. Place a 33-hole index plate onto the index (dividing) head.

11. Adjust the sector arms to 27 holes.

12. Move the knee upward .045" for the first cut.

13. Power-up the mill. Be sure that you have your eye protection on!

14. Move the table in the X-axis and cut the workpiece.

15. Move the table up another .045" for a total of .090" and back up the table to finish the cut.
16. Drop the table down to .045".

17. Rotate the index-handle clockwise 1 and $^{27}/_{33}$ of a turn.

18. Rotate the sector arms counter-clockwise until the right side arm touches the index handle. This establishes your new starting point for the next gear tooth.

19. Repeat the previous steps, 12 through 18, until all teeth are cut.

20. When completed, remove the gear and arbor assembly from the index (dividing) head. Remove balance of the tools and clean the machine for the next project.

If you did everything correctly, you have a 22-tooth timing gear in your hands. Congratulations!

Meat Tenderizer

This project will demonstrate how a double-angle cutter is mounted onto a stub arbor and used to cut a series of grooves into the surface of a workpiece to create a meat tenderizer hammerhead. Dimensions will be given for construction of the handle, but will not be covered in this text.

For this project you will be working in the Y- and Z-axes for cutter positioning, and the X-axis will be used to control depth of cut. This is not as difficult as it may first appear. We need to cut on the ends of the workpiece because the double-angle cutter must be mounted vertically.

The project will be made from a piece of 2" × 2" × 3" aluminum (alloy type not important) or plastic stock. See the drawings at the end of this chapter for details.

> **!**
> If plastic stock is used, be sure to check with your supplier for compatibility with the food service industry. Certain plastics may not be suitable for this project.

Use the following steps to machine the meat tenderizer:

1. Ends of the workpiece should be square. Layout the locations using traditional layout practices. The small X located in the upper left-hand corner is where the top peak of the pyramid shaped tooth will be positioned. Clamp workpiece into a pre-aligned vise.

2. Install 90° double-angle cutter onto a stub arbor. Install the stub arbor into the spindle. Be sure to set the length of the cutter far enough out of the spindle to enable cutting all of the tenderizer teeth.

3. This picture illustrates what can happen if this length is not correct. Set speed to 400 rpm.

4. Raise the table so that the center of the cutter is located on the top edge of the workpiece. ZERO all table feed dials.

5. After touching-off in the edge of the workpiece with the cutter, move the workpiece towards the cutter .225". Note the cutter should be positioned at the .225" measurement with the workpiece out of the way. Make the cut after you have moved the table into the correct position.

6. Make the first conventional milling cut across the face of the workpiece. Return the cutter to the backside of the workpiece. Move the knee up .500" and repeat the milling procedure. This will complete the first full tooth of the tenderizer. Again, return the cutter to the opposite side of the workpiece. Raise the knee another .500" and make the next cut. Repeat this process until all teeth are formed.

7. Next, find part zero and drill the .750" diameter through hole for the handle into the workpiece.

8. Remove the workpiece and rotate 90° and re-clamp. Center the cutter to the top edge of the workpiece. Cut a chamfer angle onto the top side. Lower the cutter 2.000" and cut the bottom chamfer.

9. Remove the workpiece, turn end-for-end and re-clamp. Repeat the steps outlined above, plus cut additional grooves to form the pyramid shaped teeth by moving the knee in .500" increments and milling.

10. This is what the tenderizer should look like after you have turned the piece 90° and milled the remaining teeth.

11. Each end of the tenderizer has a different use. Consult your handy dandy cookbook to determine which end is used for different types of meat, fish and poultry.

A piece of $^{3}/_{4}$" hardwood doweling can be used to serve as a handle for the meat tenderizer. Use a rust-proof fastener to secure the handle to the tenderizer head.

Afterword

If you have read the entire book and completed all of the projects, you are on your way to understanding the fundamentals of turret mill operation. No book can teach you every aspect of the machining process and operational procedures. However, with the application of the basics we have covered, you should have a pretty good idea of how to apply safe operating procedures, understand mill terminology, identify various cutter types, and the application of work holding techniques.

By applying the steps learned in this book and understanding new techniques and operational procedures taught to you by teachers and journeyman machinists, you will enjoy the success of having mastered a wonderful new tool.

And, one last comment—don't forget those safety glasses and safety rules taught in the very first chapter of this book.

Notes:

Project Materials and Tool List

Jet Holder

Aluminum 6061-T6
#3 center-drill
#8 drill
1/4"-28 plug tap

Adjustable Bracket

Aluminum 6061-T6
#3 center-drill
3/8" jobbers length drill bit

Bolt Pattern

#3 center-drill
1/4" jobbers length drill bit
Aluminum 6061-T6

Bolt Circle

#3 center-drill
6.00 mm jobbers length drill bit
Aluminum 6061-T6

Keyway – Slot Type

1/4" two-flute end mill
Steel 1012 CRS

Keyway – Woodruff Key Type

#201 Keyway cutter
Steel 1012 CRS

Timing Belt Gear

Acrylic plastic or Plexiglas
#3 center-drill
1/4" jobbers length drill bit
20° tapered end mill, w/.090 small-end diameter
Index (dividing) head
1/2" arbor
15/32" jobbers length drill bit
.500" straight-flute reamer
#7 drill
1/4"-20 plug tap

Meat Tenderizer

60° double-angle cutter
Stub arbor (to fit double-angle cutter)
Aluminum 6061-T6

Project Drawings

The following drawings will be useful to you when you begin your sample projects. They have information you will beed to produce projects correctly. These basic drawings are similar to those used in industry. The majority of the information provided is in the *absolute* measurement system.

You may find it useful to convert them to incremental measurements. Your instructor may require you to convert the drawings to become familiar with modifying blueprints. Modification of blueprints is a common practice. The engineer who drew the prints may not be familiar with machine operations and practices. This may require that you re-compute part lengths so that they are easier to read and follow. **Note:** If you do make modifications to the prints, be sure to have your supervisor check them for mistakes, and accuracy. Your job may depend upon it.

Use of the drawings will most certainly help when filling out process and set-up sheets for each project. The importance of using these sheets cannot be overstressed. As a matter of course, learn how to use process and set-up sheets. It will be surprising how much they will speed up your job and reduce wasted machine time trying to figure out what you should do next.

Each drawing will include the material type, material size, surface finish, specification allowances, part diagram, comment space, student name, and the date on which the project was completed and turned in for inspection.

PROJECT: Carburetor Jet Holder

Material: *6061T-6 Aluminum alloy* **Size:** *1-1/2" x 1/4"* **Surface Finish:** *Mill finish* **Specifications:** *All specifications ±.005"*

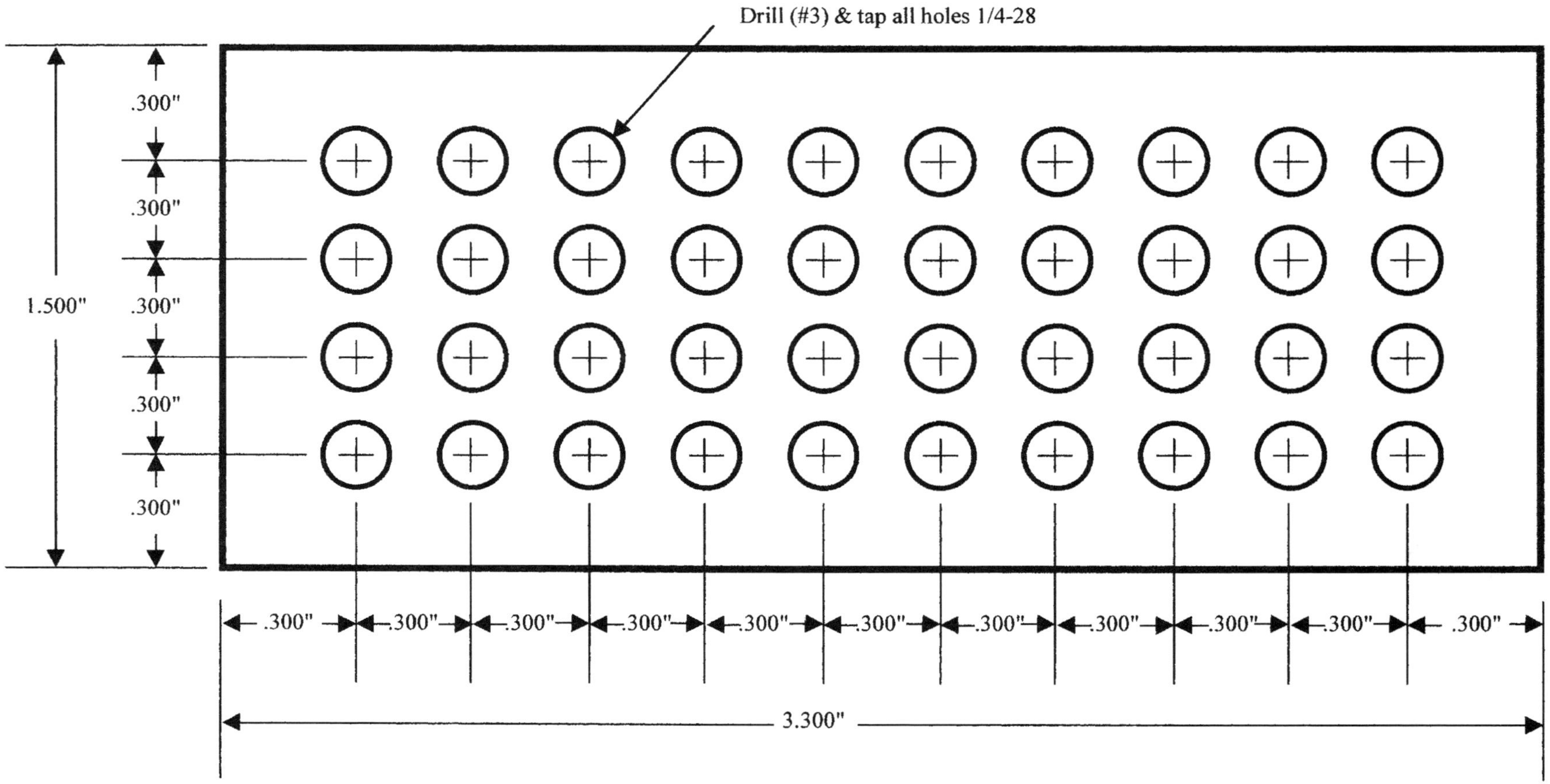

Comments:

Student: ______________________ **Date:** ___/___/___

PROJECT: Adjustable Bracket

Material: 6061T-651 Aluminum alloy **Size:** 1" x 3/8"
Surface Finish: Mill finish
Specifications: All specifications ±.005"

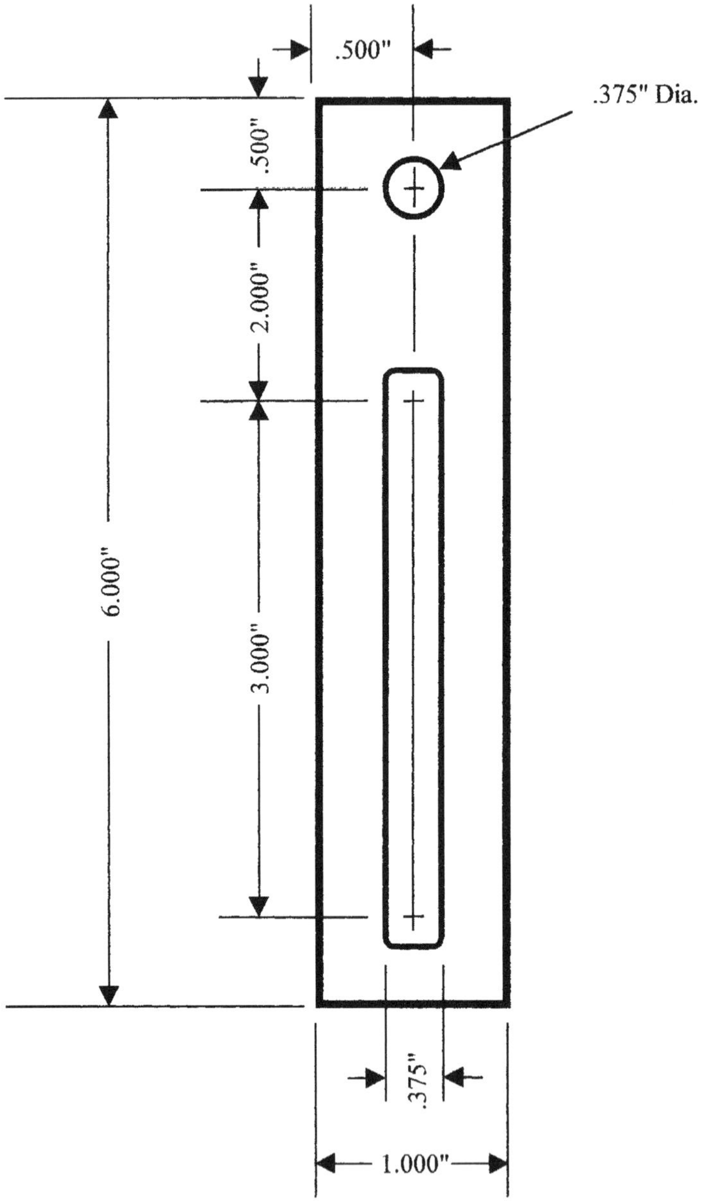

Comments:

Student: ______________________ **Date:** ___/___/___

PROJECT: Bolt Pattern

Material: *6061T-6 Aluminum alloy* **Size:** *2" x 3/8"*
Surface Finish: *Mill finish*
Specifications: *All specifications ±.005"*

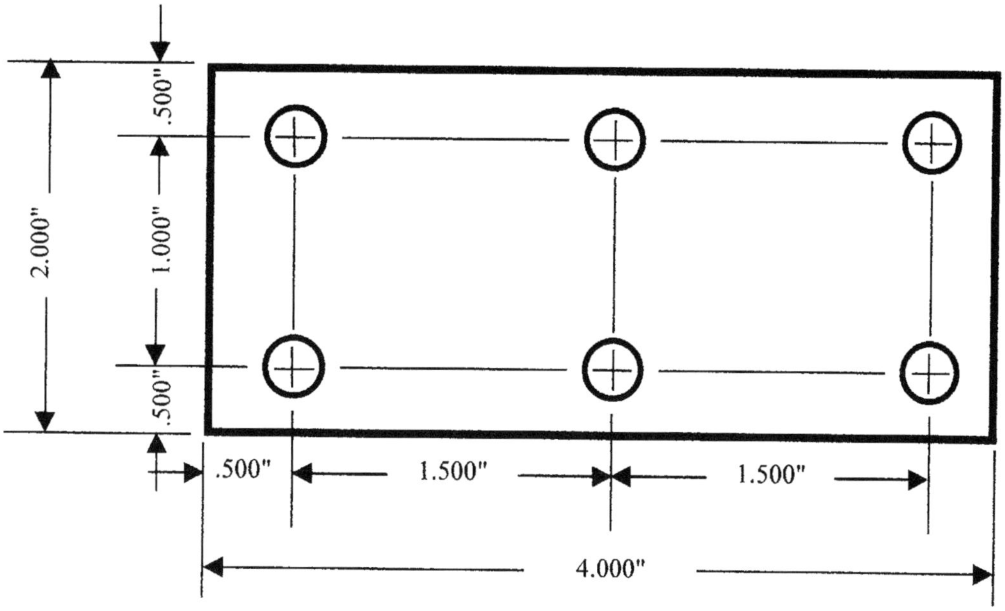

Comments:

Student: ______________________ **Date:** ___/___/___

PROJECT: Bolt Circle

Material: *6061T-651 Aluminum alloy* **Size:** *50,00 mm Dia. X 8,00 mm*

Surface Finish: *Mill finish* **Specifications:** *All specifications ±0,10 mm*

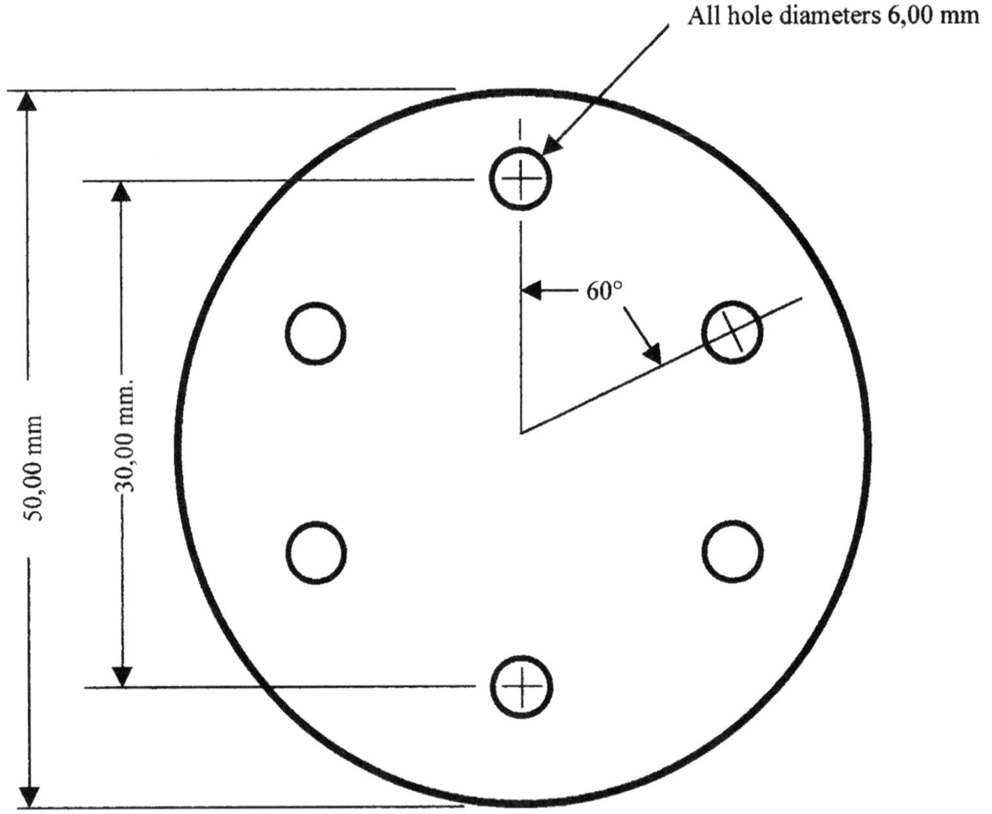

Note: 6 holes @ 60° apart

Comments:

Student: ______________________ **Date:** ___/___/___

PROJECT: Meat Tenderizer

Material: *6061T-651 Aluminum alloy* **Size:** *2" x 2"* **Surface Finish:** *Mill finish* **Specifications:** *All specifications ±.010"*

Page: 1 of 2

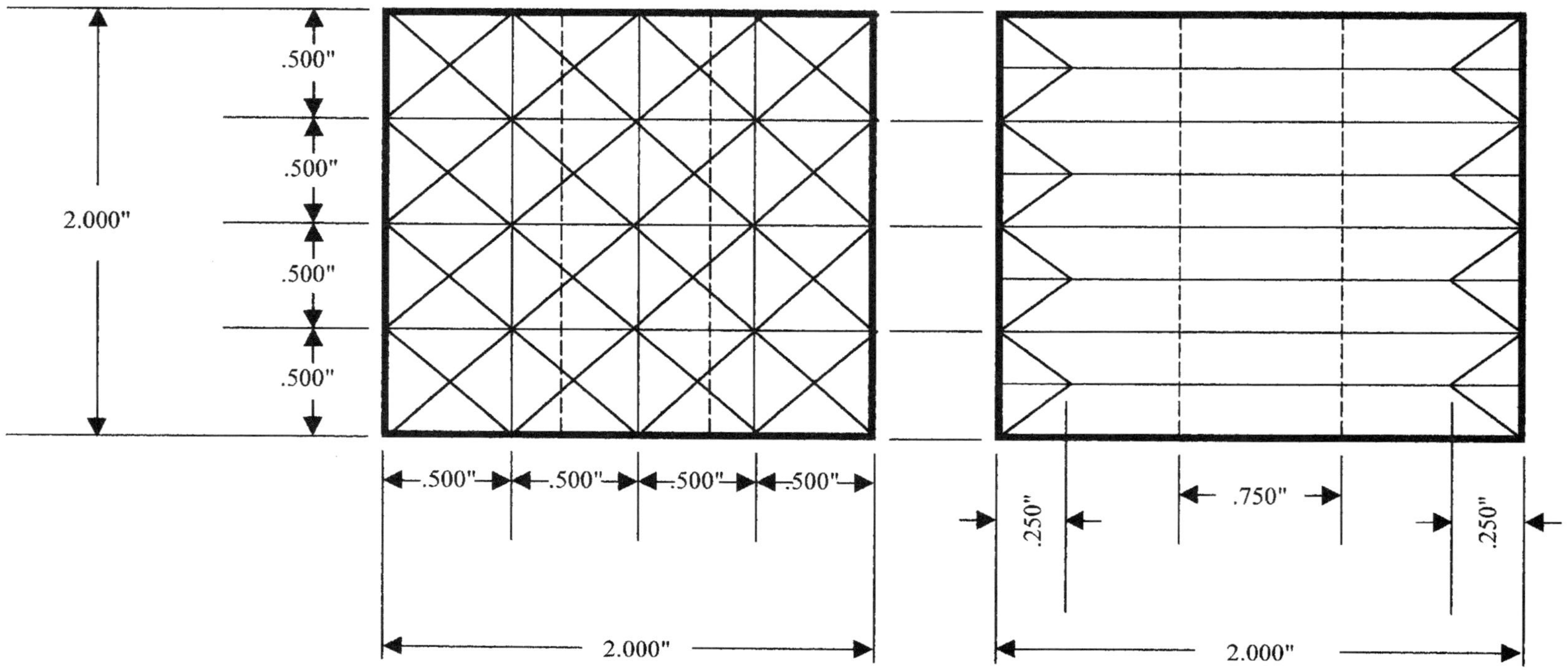

Comments:

Student: ______________________ **Date:** ___/___/___

Courtesy JO/EL Engineering Technical Publications © 1999

PROJECT: Meat Tenderizer

Material: *6061T-651 Aluminum alloy* **Size:** *2" x 2"* **Surface Finish:** *Mill finish* **Specifications:** *All specifications ±.010"*

Page: 2 of 2

.750 diameter hole must be centered (1.000") to width of part (2.000") and length (2.750")

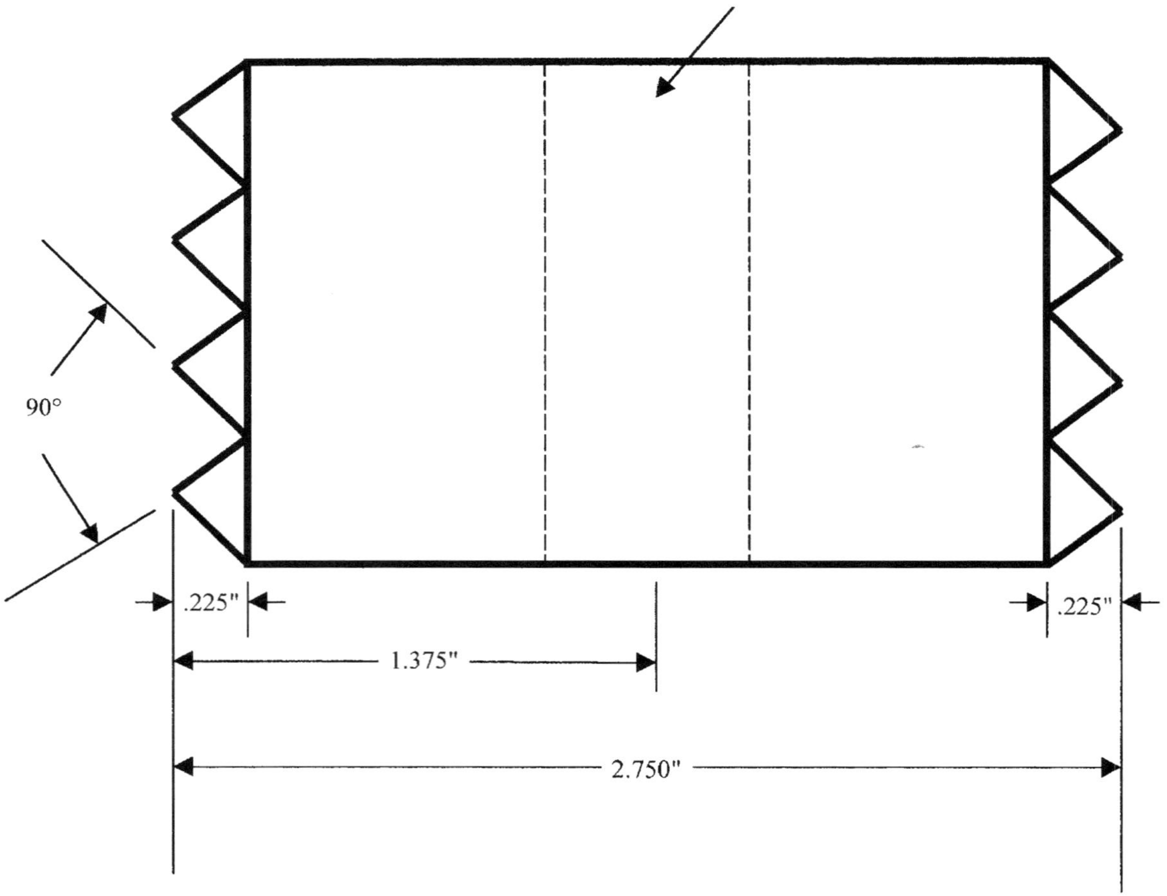

Comments:

Student: ______________________ **Date:** ___/___/___

Appendix

RIGHT ANGLE TRIANGLES

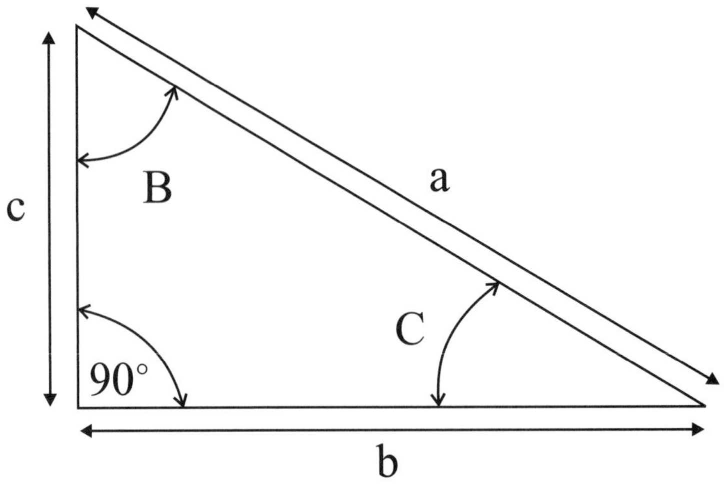

Given: the length of two sides. Find the length of the third side.

$a = \sqrt{b^2 + c^2}$ $b = \sqrt{a^2 - c^2}$ $c = \sqrt{a^2 - b^2}$

Given: The hypotenuse (a), and one acute angle B. Find b, c, and C.

$b = a \times \sin B$ $c = a \times \cos B$ $C = 90° - B$

Given: The hypotenuse (a), and one acute angle C. Find b, c, and B.

$b = a \times \cos C$ $c = a \times \sin C$ $B = 90° - C$

Given: Acute angle B and its adjacent side c. Find a, b, and C.

$a = c \div \cos B$ $b = c \times \tan B$ $C = 90° - B$

Given: Acute angle C and its adjacent side b. Find a, c, and B.

$a = b \div \cos C$ $c = b \times \tan C$ $B = 90° - C$

Given: Acute angle B and the opposite side b. Find a, c, and C.

$a = b \div \sin B$ $c = b \times \cot B$ $C = 90° - B$

Given: Acute angle C and the opposite side c. Find a, b, and B.

$a = c \div \sin C$ $b = c \times \cot C$ $B = 90° - C$

Given B: $C = 90° - B$

Given C: $B = 90° - C$

To find the **sine** of an angle, divide the side opposite by the hypotenuse.

example: $\sin C = c \div a$

To find the **cosine** of an angle, divide the side adjacent by the hypotenuse.

example: $\cos C = b \div a$

To find the **tangent** of an angle, divide the side opposite by the side adjacent.

example: $\tan C = c \div b$

To find the **cotangent** of an angle, divide the side adjacent by the side opposite.

example: $\text{cotan}\ C = b \div c$

To find the **area** of a right angle triangle.

$\text{area} = b \times c \div 2$

OBLIQUE ANGLE TRIANGLES

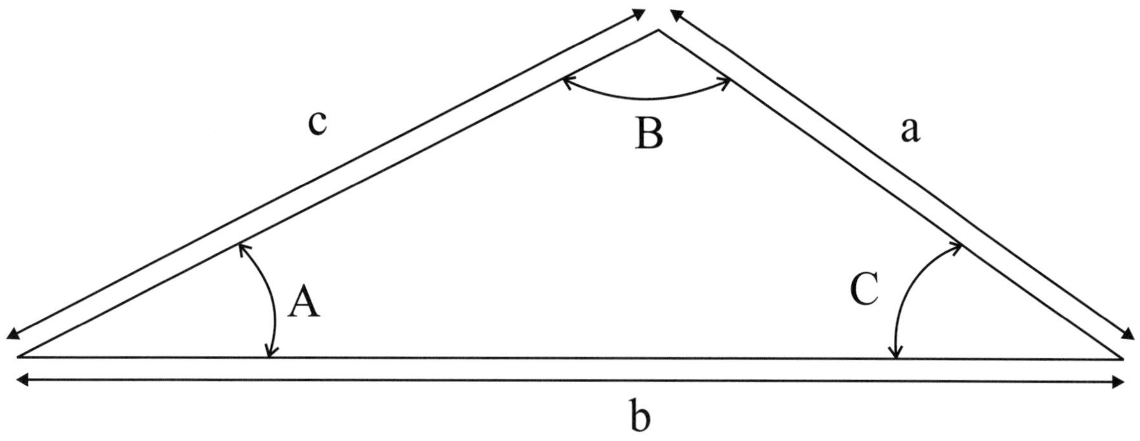

Given: One side a and two angles A and B. Find b, c, and C.

$b = (a \times \sin B) \div \sin A$ $c = (a \times \sin C) \div \sin A$ $C = 180° - (A + B)$

Note: The unknown side = the known side × the sine of the angle opposite the unknown side ÷ the sine of the angle opposite the known side.

Note: The sum of the triangle always = 180°. Therefore, when two angles are given, the third angle can be found by subtracting the sum of the given angles from 180°.

Given: Opposite angle A, and two sides a and b. Find B, C, and c.

$\sin B = (b \times \sin A) \div a$ $C = 180° - (A + B)$ $c = (a \times \sin C) \div \sin A$

Given: Sides a and b, and the angle between them, C. Find A, B, and c.

$\tan A = (a \times \sin C) \div (b - [a \times \cos C])$ $B = 180° - (A + C)$

$c = (a \times \sin C) \div \sin A$

Given: All sides, a, b, c. Find A, B, and C.

$\text{Cos } A = (b^2 + c^2 - a^2) \div (2 \times b \times c)$ $\text{Cos } B = (a^2 + c^2 - b^2) \div (2 \times a \times c)$

$\text{Cos } C = (a^2 + b^2 - c^2) \div (2 \times a \times b)$

Find the area. The area of a triangle is equal to the base × the altitude ÷ 2.

$\text{Area} = A + B + C \div 2$

When two sides and the angle between them is given.

$\text{Area} = (a \times b \times \sin C) \div 2$

TABLE OF SINES

Read degrees in left-hand column and minutes at top

Example: sin 9° 30' = .16504

Deg.	0'	10'	20'	30'	40'	50'	60'	
0	.00000	.00291	.00581	. 00872	.01163	.01454	.01745	89
1	.01745	.02036	.02326	.02617	.02908	.03199	.03489	88
2	.03489	.03780	.04071	.04361	.04652	.04943	.05233	87
3	.05233	.05524	.05814	.06104	.06395	.06685	.06975	86
4	.06975	.07265	.07555	.07845	.08135	.08425	.08715	85
5	.08715	.09005	.09295	.09584	.09874	.10163	.10452	84
6	.10452	.10742	.11031	.11320	.11609	.11898	.12186	83
7	.12186	.12475	.12764	.13052	.13341	.13629	.13917	82
8	.13917	.14205	.14493	.14780	.15068	.15356	.15643	81
9	.15643	.15930	.16217	.16504	.16791	.17078	.17364	80
10	.17364	.17651	.17937	.18223	.18509	.18795	.19080	79
11	.19080	.19366	.19651	.19936	.20221	.20506	.20791	78
12	.20791	.21075	.21359	.21644	.21927	.22211	.22495	77
13	.22495	.22778	.23061	.23344	.23627	.23909	.24192	76
14	.24192	.24474	.24756	.25038	.25319	.25600	.25881	75
15	.25881	.26162	.26443	.26723	.27004	.27284	.27563	74
16	.27563	.27843	.28122	.28401	.28680	.28958	.29237	73
17	.29237	.29515	.29793	.30070	.30347	.30624	.30901	72
18	.30901	.31178	.31454	.31730	.32006	.32281	.32556	71
19	.32556	.32831	.33106	.33380	.33654	.33928	.34202	70
20	.34202	.34475	.34748	.35021	.35293	.35565	.35836	69
21	.35836	.36108	.36379	.36650	.36920	.37190	.37460	68
22	.37460	.37730	.37999	.38268	.38536	.38805	.39073	67
23	.39073	.39340	.39607	.39874	.40141	.40407	.40673	66
24	.40673	.40939	.41204	.41469	.41733	.41998	.42261	65
25	.42261	.42525	.42788	.43051	.43313	.43575	.43837	64
26	.43837	.44098	.44359	.44619	.44879	.45139	.45399	63
27	.45399	.45658	.45916	.46174	.46432	.46690	.46947	62
28	.46947	.47203	.47460	.47715	.47971	.48226	.48481	61
29	.48481	.48735	.48989	.49242	.49495	.49747	.50000	60
30	.50000	.50251	.50503	.50753	.51004	.51254	.51504	59
31	.51504	.51752	.52001	.52249	.52497	.52745	.52991	58
32	.52991	.53238	.53484	.53730	.53975	.54219	.54463	57
33	.54463	.54707	.54950	.55193	.55436	.55677	.55919	56
34	.55919	.56160	.56400	.56640	.56880	.57119	.57358	55
35	.57358	.57595	.57833	.58070	.58306	.58542	.58778	54
36	.58778	.59013	.59248	.59482	.59715	.59948	.60181	53
37	.60181	.60413	.60645	.60876	.61106	.61336	.61566	52
38	.61566	.61795	.62023	.62251	.62478	.62705	.62932	51
39	.62932	.63157	.63383	.63607	.63832	.64055	.64279	50
40	.64279	.64501	.64723	.64944	.65165	.65386	.65605	49
41	.65605	.65825	.66043	.66262	.66479	.66696	.66913	48
42	.66913	.67128	.67344	.67559	.67773	.67986	.68199	47
43	.68199	.68412	.68624	.68835	.69046	.69256	.69465	46
44	.69465	.69674	.69883	.70090	.70298	.70504	.70710	45
	60'	50'	40'	30'	20'	10'	0'	Deg.

TABLE OF COSINES

Read degrees in right-hand column and minutes at bottom

Example: cos 52° 10' = .61336

TABLE OF SINES

Read degrees in left-hand column and minutes at top
Example: sin 50° 40' = .77347

Deg.	0'	10'	20'	30'	40'	50'	60'	
45	.70710	.70916	.71120	.71325	.71528	.71731	.71934	44
46	.71934	72135	.72336	.72537	.72737	.72936	.73135	43
47	.73135	.73333	.73530	.73727	.73923	.74119	.74314	42
48	.74314	.74508	.74702	.74895	.75088	.75279	.75471	41
49	.75471	.75661	.75851	.76040	.76229	.76417	.76604	40
50	.76604	.76791	.76977	.77162	.77347	.77531	.77714	39
51	.77714	.77897	.78079	.78260	.78441	.78621	.78801	38
52	.78801	.78979	.79157	.79335	.79512	.79688	.79863	37
53	.79863	.80038	.80212	.80385	.80558	.80730	.80901	36
54	.80901	.81072	.81242	.81411	.81580	.81748	.81915	35
55	.81915	.82081	.82247	.82412	.82577	.82740	.82903	34
56	.82903	.83066	.83227	.83388	.83548	.83708	.83867	33
57	.83867	.84025	.84182	.84339	.84495	.84650	.84804	32
58	.84804	.84958	.85111	.85264	.85415	.85566	.85717	31
59	.85717	.85866	.86014	.86162	.86310	.86457	.86602	30
60	.86602	.86747	.86892	.87035	.87178	.87320	.87462	29
61	.87462	.87602	.87742	.87881	.88020	.88157	.88295	28
62	.88295	.88430	.88566	.88701	.88835	.88968	.89100	27
63	.89100	.89232	.89363	.89493	.89622	.89751	.89879	26
64	.89879	.90006	.90132	.90258	.90383	.90507	.90630	25
65	.90630	.90753	.90875	.90996	.91116	.91235	.91354	24
66	.91354	.91472	.91589	.91706	.91821	.91936	.92050	23
67	.92050	.92163	.92276	.92388	.92498	.92609	.92718	22
68	.92718	.92827	.92934	.93041	.93148	.93253	.93358	21
69	.93358	.93461	.93565	.93667	.93768	.93869	.93969	20
70	.93969	.94068	.94166	.94264	.94360	.94456	.94551	19
71	.94551	.94646	.94739	.94832	.94924	.95015	.95105	18
72	.95105	.95195	.95283	.95371	.95458	.95545	.95630	17
73	.95630	.95715	.95799	.95882	.95964	.96045	.96126	16
74	.96126	.96205	.96284	.96363	.96440	.96516	.96592	15
75	.96592	.96667	.96741	.96814	.96887	.96958	.97029	14
76	.97029	.97099	.97168	.97237	.97304	.97371	.97437	13
77	.97437	.97502	.97566	.97629	.97692	.97753	.97815	12
78	.97815	.97874	.97934	.97992	.98050	.98106	.98162	11
79	.98162	.98217	.98272	.98325	.98378	.98429	.98481	10
80	.98481	.98530	.98580	.98628	.98676	.98722	.98768	9
81	.98768	.98813	.98858	.98901	.98944	.98985	.99026	8
82	.99026	.99066	.99106	.99144	.99182	.99218	.99254	7
83	.90254	.99289	.99323	.99357	.99389	.99421	.99452	6
84	.99452	.99482	.99511	.99539	.99567	.99593	.99619	5
85	.99619	.99644	.99668	.99691	.99714	.99735	.99756	4
86	.99756	.99776	.99795	.99813	.99830	.99847	.99863	3
87	.99863	.99877	.99891	.99904	.99917	.99928	.99939	2
88	.99939	.99948	.99957	.99965	.99972	.99979	.99984	1
89	.99984	.99989	.99993	.99996	.99998	.99999	1.00000	0
	60'	50'	40'	30'	20'	10'	0'	Deg.

TABLE OF COSINES

Read degrees in right-hand column and minutes at bottom
Example: cos 11° 20' = .98050

TABLE OF TANGENTS

Read degrees in left-hand column and minutes at top
Example: tan 15° 20' = .27419

Deg.	0'	10'	20'	30'	40'	50'	60'	
0	.00000	.00290	.00581	.00872	.01163	.01454	.01745	89
1	.01745	.02036	.02327	.02618	.02909	.03200	.03492	88
2	.03492	.03783	.04074	.04366	.04657	.04949	.05240	87
3	.05240	.05532	.05824	.06116	.06408	.06700	.06992	86
4	.06992	.07285	.07577	.07870	.08162	.08455	.08748	85
5	.08748	.09042	.09335	.09628	.09922	.10216	.10510	84
6	.10510	.10804	.11099	.11393	.11688	.11983	.12278	83
7	.12278	.12573	.12869	.13165	.13461	.13757	.14054	82
8	.14054	.14350	.14647	.14945	.15242	.15540	.15838	81
9	.15838	.16136	.16435	.16734	.17033	.17332	.17632	80
10	.17632	.17932	.18233	.18533	.18834	.19136	.19438	79
11	.19438	.19740	.20042	.20345	.20648	.20951	.21255	78
12	.21255	.21559	.21864	.22169	.22474	.22780	.23086	77
13	.23086	.23393	.23700	.24007	.24315	.24624	.24932	76
14	.24932	.25242	.25551	.25861	.26172	.26483	.26794	75
15	.26794	.27106	.27419	.27732	.28046	.28360	.28674	74
16	.28674	.28989	.29305	.29621	.29938	.30255	.30573	73
17	.30573	.30891	.31210	.31529	.31850	.32170	.32492	72
18	.32492	.32813	.33136	.33459	.33783	.34107	.34432	71
19	.34432	.34758	.35084	.35411	.35739	.36067	.36397	70
20	.36397	.36726	.37057	.37388	.37720	.38053	.38386	69
21	.38386	.38720	.39055	.39391	.39727	.40064	.40402	68
22	.40402	.40741	.41080	.41421	.41762	.42104	.42447	67
23	.42447	.42791	.43135	.43481	.43827	.44174	.44522	66
24	.44522	.44871	.45221	.45572	.45924	.46277	.46630	65
25	.46630	.46985	.47341	.47697	.48055	.48413	.48773	64
26	.48773	.49133	.49495	.49858	.50221	.50586	.50952	63
27	.50952	.51319	.51687	.52056	.52427	.52798	.53170	62
28	.53170	.53544	.53919	.54295	.54672	.55051	.55430	61
29	.55430	.55811	.56193	.56577	.56961	.57347	.57735	60
30	.57735	.58123	.58513	.58904	.59297	.59690	.60086	59
31	.60086	.60482	.60880	.61280	.61680	.62083	.62486	58
32	.62486	.62892	.63298	.63707	.64116	.64528	.64940	57
33	.64940	.65355	.65771	.66188	.66607	.67028	.67450	56
34	.67450	.67874	.68300	.68728	.69157	.69588	.70020	55
35	.70020	.70455	.70891	.71329	.71769	.72210	.72654	54
36	.72654	.73099	.73546	.73996	.74447	.74900	.75355	53
37	.75355	.75812	.76271	.76732	.77195	.77661	.78128	52
38	.78128	.78598	.79069	.79543	.80019	.80497	.80978	51
39	.80978	.81461	.81946	.82433	.82923	.83415	.83910	50
40	.83910	.84406	.84906	.85408	.85912	.86419	.86928	49
41	.86928	.87440	.87955	.88472	.88992	.89515	.90040	48
42	.90040	.90568	.91099	.91633	.92169	.92709	.93251	47
43	.93251	.93796	.94345	.94896	.95450	.96008	.96568	46
44	.96568	.97132	.97699	.98269	.98843	.99419	1.00000	43
	60'	50'	40'	30'	20'	10'	0'	Deg.

TABLE OF COTANGENTS

Read degrees in right-hand column and minutes at bottom
Example: cot 47° 50' = .90568

TABLE OF TANGENTS

Read degrees in left-hand column and minutes at top
Example: tan 47° 30' = 1.0913

Deg.	0'	10'	20'	30'	40'	50'	60'	
45	1.0000	1.0058	1.0117	1.0176	1.0235	1.0295	1.0355	44
46	1.0355	1.0415	1.0476	1.0537	1.0599	1.0661	1.0723	43
47	1.0723	1.0786	1.0849	1.0913	1.0977	1.1041	1.1106	42
48	1.1106	1.1171	1.1236	1.1302	1.1369	1.1436	1.1503	41
49	1.1503	1.1571	1.1639	1.1708	1.1777	1.1847	1.1917	40
50	1.1917	1.1988	1.2059	1.2131	1.2203	1.2275	1.2349	39
51	1.2349	1.2422	1.2496	1.2571	1.2647	1.2723	1.2799	38
52	1.2799	1.2876	1.2954	1.3032	1.3111	1.3190	1.3270	37
53	1.3270	1.3351	1.3432	1.3514	1.3596	1.3680	1.3763	36
54	1.3763	1.3848	1.3933	1.4019	1.4106	1.4193	1.4281	35
55	1.4281	1.4370	1.4459	1.4550	1.4641	1.4733	1.4825	34
56	1.4825	1.4919	1.5013	1.5108	1.5204	1.5301	1.5398	33
57	1.5398	1.5497	1.5596	1.5696	1.5798	1.5900	1.6003	32
58	1.6003	1.6107	1.6212	1.6318	1.6425	1.6533	1.6642	31
59	1.6642	1.6753	1.6864	1.6976	1.7090	1.7204	1.7320	30
60	1.7320	1.7437	1.7555	1.7674	1.7795	1.7917	1.8040	29
61	1.8040	1.8164	1.8290	1.8417	1.8546	1.8676	1.8807	28
62	1.8807	1.8940	1.9074	1.9209	1.9347	1.9485	1.9626	27
63	1.9626	1.9768	1.9911	2.0056	2.0203	2.0352	2.0503	26
64	2.0503	2.0655	2.0809	2.0965	2.1123	2.1283	2.1445	25
65	2.1445	2.1609	2.1774	2.1943	2.2113	2.2285	2.2460	24
66	2.2460	2.2637	2.2816	2.2998	2.3182	2.3369	2.3558	23
67	2.3558	2.3750	2.3944	2.4142	2.4342	2.4545	2.4750	22
68	2.4750	2.4959	2.5171	2.5386	2.5604	2.5826	2.6050	21
69	2.6050	2.6279	2.6510	2.6746	2.6985	2.7228	2.7474	20
70	2.7474	2.7725	2.7980	2.8239	2.8502	2.8770	2.9042	19
71	2.9042	2.9318	2.9600	2.9886	3.0178	3.0474	3.0776	18
72	3.0776	3.1084	3.1397	3.1715	3.2040	3.2371	3.2708	17
73	3.2708	3.3052	3.3402	3.3759	3.4123	3.4495	3.4874	16
74	3.4874	3.5260	3.5655	3.6058	3.6470	3.6890	3.7320	15
75	3.7320	3.7759	3.8208	3.8667	3.9136	3.9616	4.0107	14
76	4.0107	4.0610	4.1125	4.1653	4.2193	4.2747	4.3314	13
77	4.3314	4.3896	4.4494	4.5107	4.5736	4.6382	4.7046	12
78	4.7046	4.7728	4.8430	4.9151	4.9894	5.0658	5.1445	11
79	5.1445	5.2256	5.3092	5.3955	5.4845	5.5763	5.6712	10
80	5.6712	5.7693	5.8708	5.9757	6.0844	6.1970	6.3137	9
81	6.3137	6.4348	6.5605	6.6911	6.8269	6.9682	7.1153	8
82	7.1153	7.2687	7.4287	7.5957	7.7703	7.9530	8.1443	7
83	8.1443	8.3449	8.5555	8.7768	9.0098	9.2553	9.5143	6
84	9.5143	9.7881	10.078	10.385	10.711	11.059	11.430	5
85	11.430	11.826	12.250	12.706	13.196	13.726	14.300	4
86	14.300	14.924	15.604	16.349	17.169	18.075	19.081	3
87	19.081	20.205	21.470	22.904	24.541	26.431	28.636	2
88	28.636	31.241	34.367	38.188	42.964	49.103	57.290	1
89	57.290	68.750	85.939	114.58	171.88	343.77	∞	0
	60'	50'	40'	30'	20'	10'	0'	Deg.

TABLE OF COTANGENTS

Read degrees in right-hand column and minutes at bottom
Example: cot 5° 10' = 11.059

The Metric System

1 inch = 25.4 millimeters
1 millimeter = 0.03937 inch
1 square millimeter = 0.00155 square inch
10 millimeters = 1 centimeter
1 centimeter = 0.3937 inch
1 square centimeter = 0.155 square inch

1 foot = 304.8 millimeters
1 yard = 0.9144 meter
1 meter = 39.37 inches
1 pound = 0.4536 kilogram
1 quart = 0.946 liter
1 gallon = 3.785 liters

Millimeters	Inches	Millimeters	Inches	Millimeters	Inches
0.25 mm	0.0098"	10.5 mm	0.4134"	279.4 mm	11.0"
0.50	0.0197	11.0	0.4331	304.8	12.0
0.75	0.0295	11.5	0.4528	330.2	13.0
1.0	0.0394	12.0	0.4724	355.6	14.0
1.25	0.0492	12.5	0.4921	381.0	15.0
1.50	0.0591	13.0	0.5118	406.4	16.0
1.75	0.0689	13.5	0.5315	431.8	17.0
2.0	0.0787	14.0	0.5512	457.2	18.0
2.25	0.0886	14.5	0.5709	482.6	19.0
2.50	0.0984	15.0	0.5906	508.0	20.0
2.75	0.1083	15.5	0.6102	533.4	21.0
3.0	0.1181	16.0	0.6299	558.8	22.0
3.25	0.1280	16.5	0.6496	584.2	23.0
3.50	0.1378	17.0	0.6693	609.6	24.0
3.75	0.1476	17.5	0.6890	635.0	25.0
4.0	0.1575	18.0	0.7087	660.4	26.0
4.25	0.1673	18.5	0.7283	685.8	27.0
4.50	0.1772	19.0	0.7480	711.2	28.0
4.75	0.1870	19.5	0.7677	736.6	29.0
5.0	0.1969	20.0	0.7874	762.0	30.0
5.25	0.2067	20.5	0.8071	787.4	31.0
5.50	0.2165	21.0	0.8268	812.8	32.0
5.75	0.2264	21.5	0.8465	838.2	33.0
6.0	0.2362	22.0	0.8661	863.6	34.0
6.25	0.2461	22.5	0.8858	889.0	35.0
6.50	0.2559	23.0	0.9055	914.4	36.0
6.75	0.2657	23.5	0.9252	939.8	37.0
7.0	0.2756	24.0	0.9449	965.2	38.0
7.25	0.2854	24.5	0.9646	990.6	39.0
7.50	0.2953	25.0	0.9843	1016.0	40.0
7.75	0.3051	25.4	1.0	1041.4	41.0
8.0	0.3150	50.8	2.0	1066.8	42.0
8.25	0.3248	76.2	3.0	1092.2	43.0
8.50	0.3346	101.6	4.0	1117.6	44.0
8.75	0.3445	127.0	5.0	1143.0	45.0
9.0	0.3543	152.4	6.0	1168.4	46.0
9.25	0.3642	177.8	7.0	1193.8	47.0
9.50	0.3740	203.2	8.0	1219.2	48.0
9.75	3.3839	228.6	9.0	1244.6	49.0
10.0	0.3937	254.0	10.0	1270.0	50.0

Inches		Millimeters	Inches		Millimeters
Decimal*	Fraction		Decimal*	Fraction	
0.0156	$^{1}/_{64}$	0.3968	0.5156	$^{33}/_{64}$	13.0966
0.0312	$^{1}/_{32}$	0.7937	0.5312	$^{17}/_{32}$	13.4934
0.0469	$^{3}/_{64}$	1.1906	0.5469	$^{35}/_{64}$	13.8903
0.0625	$^{1}/_{16}$	1.5875	0.5625	$^{9}/_{16}$	14.2872
0.0781	$^{5}/_{64}$	1.9843	0.5781	$^{37}/_{64}$	14.6841
0.0938	$^{3}/_{32}$	2.3812	0.5938	$^{19}/_{32}$	15.0809
0.1094	$^{7}/_{64}$	2.7780	0.6094	$^{39}/_{64}$	15.4778
0.1250	$^{1}/_{8}$	3.1749	0.6250	$^{5}/_{8}$	15.8747
0.1406	$^{9}/_{64}$	3.5718	0.6406	$^{41}/_{64}$	16.2715
0.1562	$^{5}/_{32}$	3.9686	0.6562	$^{21}/_{32}$	16.6684
0.1719	$^{11}/_{64}$	4.3655	0.6719	$^{43}/_{64}$	17.0653
0.1875	$^{3}/_{16}$	4.7624	0.6875	$^{11}/_{16}$	17.4621
0.2031	$^{13}/_{64}$	5.1592	0.7031	$^{45}/_{64}$	17.8590
0.2188	$^{7}/_{32}$	5.5561	0.7188	$^{23}/_{32}$	18.2559
0.2344	$^{15}/_{64}$	5.9530	0.7344	$^{47}/_{64}$	18.6527
0.250	$^{1}/_{4}$	6.3498	0.750	$^{3}/_{4}$	19.0496
0.2656	$^{17}/_{64}$	6.7467	0.7656	$^{49}/_{64}$	19.4465
0.2812	$^{9}/_{32}$	7.1436	0.7812	$^{25}/_{32}$	19.8433
0.2969	$^{19}/_{64}$	7.5404	0.7969	$^{51}/_{64}$	20.2402
0.3125	$^{5}/_{16}$	7.9373	0.8125	$^{12}/_{16}$	20.6371
0.3281	$^{21}/_{64}$	8.3342	0.8281	$^{53}/_{64}$	21.0339
0.3438	$^{11}/_{32}$	8.7310	0.8438	$^{27}/_{32}$	21.4308
0.3594	$^{23}/_{64}$	9.1279	0.8594	$^{55}/_{64}$	21.8277
0.3750	$^{3}/_{8}$	9.5248	0.8750	$^{7}/_{8}$	22.2245
0.3906	$^{25}/_{64}$	9.9216	0.8906	$^{57}/_{64}$	22.6214
0.4062	$^{13}/_{32}$	10.3185	0.9062	$^{29}/_{32}$	23.0183
0.4219	$^{27}/_{64}$	10.7154	0.9219	$^{59}/_{64}$	23.4151
0.4375	$^{7}/_{16}$	11.1122	0.9375	$^{15}/_{16}$	23.8120
0.4531	$^{29}/_{64}$	11.5091	0.9531	$^{61}/_{64}$	24.2089
0.4688	$^{15}/_{32}$	11.9060	0.9688	$^{31}/_{32}$	24.6057
0.4844	$^{31}/_{64}$	12.3029	0.9844	$^{63}/_{64}$	25.0026
0.50	$^{1}/_{2}$	12.7000	1.0		25.4000

* Decimals rounded to four places.

Inch to millimeter conversions for shallow depths of cut.					
Inches	Millimeters	Inches	Millimeters	Inches	Millimeters
0.001"	0.0254 mm	0.016"	0.4064 mm	0.031"	0.7874 mm
0.002	0.0508	0.017	0.4318	0.032	0.8128
0.003	0.0762	0.018	0.4572	0.033	0.8382
0.004	0.1016	0.019	0.4826	0.034	0.8636
0.005	0.1270	0.020	0.5080	0.035	0.8890
0.006	0.1524	0.021	0.5334	0.036	0.9144
0.007	0.1778	0.022	0.5588	0.037	0.9398
0.008	0.2032	0.023	0.5842	0.038	0.9652
0.009	0.2286	0.024	0.6096	0.039	0.9906
0.010	0.2540	0.025	0.6350	0.040	1.0160
0.011	0.2794	0.026	0.6604	0.041	1.0414
0.012	0.3048	0.027	0.6858	0.042	1.0668
0.013	0.3302	0.028	0.7112	0.043	1.0922
0.014	0.3556	0.029	0.7366	0.044	1.1176
0.015	0.3810	0.030	0.7620	0.045	1.1430

Terms & Vocabulary

A

angle
The inclination of two diverging lines originating from a common point measured in degrees from 0° to 360°.

angle indexing
The positioning of the workpiece in relationship to an angle that is relative to a starting point.

Ångström unit
The unit employed for expressing wavelengths of light.

arc
A continuous portion of a curved line, as in an eclipse or segment of a circle.

automatic feed
A powered feed unit used to provide motive force to the mill table or spindle.

B

bar stock
Metal stock, having a uniform cross-section width and height, manufactured by rolling or extruding.

billet
Metal stock that was produced by rolling.

blank
A piece of stock machined to a specific size or diameter.

blind hole
A hole which does not connect two parallel surfaces.

Bridgeport® mill
The originator of the knee-type turret milling machine. Also, a term used to describe all knee-type turret mills, regardless of manufacturer.

C

calibration
A process where a tool or machine is adjusted to move in a prescribed manner, such as length or angle.

carbide
A very pure form of powdered tungsten that is pressed into a particular shape and sintered to a hardness over 90 Rockwell (C-scale) and used as a cutting tool.

CBN (cubic boron nitride)
A man-made material composed of special compressed and sintered chemical powders. CBN is very resistant to wear and heat. When used in cutting tools, it is second only to diamond (see PCD) in hardness.

chip load
The amount of stock each cutter blade removes from the workpiece.

CNC (computer numerical control) mill
A type of mill that uses computer driven motors to move the table or workpiece in X-, Y-, and Z-axis.

D

depth of cut
The thickness of the material being removed from the workpiece by the cutter blade.

detent
A ball or spring used to temporally locate, or position, a movable part of component of an assembly.

die
A cutting tool used to cut external threads onto a workpiece.

drip lubrication
A type of lubrication where coolant or lubricating oil os fed to a workpiece from a reservoir.

drive dogs
A pair of small metal blocks affixed to a tool holder to ensure a positive locking mechanism used to drive a cutter assembly.

DRO (digital readout)
An electro-mechanical device used to indicate linear movement of the workpiece in relationship to the cutting tool centerline. Distance is displayed in a digital format.

Dykem
A trade name used to describe a type of bluing ink or dye that is applied to the surface of a workpiece for layout purposes.

E

elongation
A term used to describe the increase of length of a material due to opposite (pulling) forces.

F

finish
A term used to describe the surface texture of a machined part. Typically specified in R_a, R_k, R_{ma}, and W_t parameters.

finish cut
The final machining performed on the workpiece, typically removing a few thousandths of stock at a higher than normal spindle speed.

first article
A term used to describe the first part produced from a production run. This part should be within all tolerances specified by the blueprint.

5-C collet
A machine tool collet used to hold work concentric to the machine or fixture centerline. Available with round, square, and hexagon holes.

form tool
A tool bit that is ground into a specific shape or form that may include straight, angle, tapered, or rounded cutting edges.

H

high speed steel
A tool steel used for cutting tools. May contain small percentages of cobalt and/or molybdenum to improve tool durability.

I

in calibration
A term used to describe tools that are calibrated to exact standards.

in-house
A process or operation that is performed within the confines of a single machine shop.

inspector
A person who inspects finished workpieces for size and fit.

ISO (International Standards Organization)
The international body that sets standards and procedures for measurements and machining practices.

ISO 9000
A defined manner by which a part is manufactured. A paper-trail is generated from the initial purchase of material, defined steps to produce the part, records of inspection, and final completion of the piece.

K

keyway
A slot that is cut into a shaft into which a square key is placed. Used to prevent slippage of a pulley or gear on a rotating shaft.

knee
That part of a machine that carries the work table.

P

part zero
The starting point and/or location from where the part X, Y, and Z locations start.

PCD (poly-crystalline-diamond)
A diamond coating applied to the cutting edge of the tool bit.

power feed
An electrically or mechanically powered mechanism used to move the table or quill of the mill.

profilometer
A measuring instrument used to measure surface finish. Typically reads R_t or R_{ma} values.

Q

quill
That part of the mill which moves the spindle in and out of the mill head.

R

reamer
A multi-fluted cutter used to remove a minimal amount of material from a workpiece and produce a very precise bore dimension.

R-8 collet
The collet system used on the knee-mill used to hold cutting tools.

Rockwell hardness testing
A testing procedure used to determine the surface hardness of a part. A penetrator is pushed into the surface of the test piece. Hardness is dependent upon the depth of the penetrator. Scales include, A, B, and C. Hardness can also be measured on Brinell and Vickers scales.

roughing
A term used to describe the rapid removal of material without regard for surface texture and finish.

S

SAE (Society of Automotive Engineers)
One of several American bodies that sets standards and establishes procedures for materials and manufacturing.

simple indexing
The positioning of a workpiece to any one of a number of preset locations/positions. e.g. 180°, 120°, 90°, 60°, etc.

SPC (statistical process control)
A process where part sizes are recorded to ensure that size and quality are maintained. Often used by machinists to predict when a cutting tool will need to be changed, and track how many parts can be produced before tool change-out.

spindle
The part of the milling machine that rotates the cutting tool.

T

through hole
A hole that connects two parallel surfaces.

TiCN
Titanium carbonitride. A vapor deposition coating that is used to increase the life of a cutting tool.

TiN
Titanium Nitride. A vapor deposition coating that is used to increase the life of a cutting tool.

Travadial®
A mechanical analog device used to indicate linear movement of the workpiece in relationship to the cutting tool centerline. Distance is displayed in an analog format.

trig-out
A mathematical procedure where trigonometric functions are used to calculate angles or positions relative to a fixed point.

W

Weldon shank
A type of shank used to hold most end mills into an end mill holder. A small relief is cut into the top or middle of the shank. This relief allows the set screw to clamp the end mill in place without distorting the surface of the shank.

X

X-axis
Movement of the mill table from side-to-side.

Y

Y-axis
Movement of the mill table from front-to-rear.

Z

Z-axis
Movement of the mill table or quill, up and down.

ZERO
The starting point of any numbering system. ZERO has no value.

ANSWERS TO QUESTIONS

Chapter 1 - Study Question Answers

1. C
2. B
3. C
4. D
5. D
6. D
7. D
8. C
9. C

Chapter 2 - Study Question Answers

1. D
2. C
3. D
4. C
5. C
6. A
7. B
8. C
9. D
10. C

Chapter 3 - Study Question Answers

1. C
2. A
3. B
4. B
5. A
6. D
7. C
8. D
9. C
10. B

Chapter 4 - Study Question Answers

1. C
2. D
3. B
4. A
5. C
6. A
7. B
8. A
9. C
10. C

Chapter 5 - Study Question Answers

1. B
2. C
3. D
4. B
5. C
6. B
7. C
8. C
9. C
10. 667 rpm
11. 1449 rpm
12. 2316 rpm
13. 1273 rpm
14. 1445 rpm
15. 111 rpm
16. 103 rpm
17. 8960 rpm
18. 637 rpm
19. 1455 rpm
20. 3840 rpm
21. 15 rpm
22. 2.9 rpm

Chapter 6 - Study Question Answers

1. B
2. D
3. C
4. A
5. B
6. D
7. C
8. A
9. C

Chapter 7 - Study Question Answers

1. B
2. A
3. B
4. C
5. A
6. C
7. C
8. C
9. C
10. B

Chapter 8 - Study Question Answers

1. B
2. D
3. B
4. A
5. C
6. C
7. C
8. B
9. D
10. C

Chapter 9 - Study Question Answers

1. C
2. D
3. A
4. C
5. C
6. C
7. B
8. B
9. B
10. B

Chapter 10 - Study Question Answers

1. C
2. Reduce wasted time, improve accuracy.
3. Tools, machines, and operational steps.
4. T-1, T-2, T-3, T-4, etc. Numbers.
5. A
6. B
7. D
8. The first part from a production run.
9. True

INDEX